Make:

Edible Inventions

3/17-2x

9/16

Make:
Edible Inventions

COOKING HACKS AND YUMMY RECIPES YOU CAN BUILD, MIX, BAKE, AND GROW

By Kathy Ceceri

MAKER MEDIA
SAN FRANCISCO, CA

Edible Inventions
Cooking Hacks and Yummy Recipes You Can Build, Mix, Bake, and Grow

By Kathy Ceceri
Copyright © 2016 Kathy Ceceri. All rights reserved.

Printed in Canada

Published by
Maker Media, Inc.,
1160 Battery Street East, Suite 125,
San Francisco, California 94111.

Maker Media books may be purchased for educational, business, or sales promotional use. Online editions are also available for most titles (safaribooksonline.com). For more information, contact our corporate/institutional sales department: 800-998-9938 or corporate@oreilly.com.

Publisher: Roger Stewart
Editor: Patrick Di Justo
Copy Editor: Elizabeth Campbell, Happenstance Type-O-Rama
Proofreader: Elizabeth Welch, Happenstance Type-O-Rama
Interior Designer and Compositor: Maureen Forys, Happenstance Type-O-Rama
Cover Designer: Maureen Forys, Happenstance Type-O-Rama
Indexer: Valerie Perry, Happenstance Type-O-Rama

September 2016: First Edition
Revision History for the First Edition
2016-09-15 First Release

See oreilly.com/catalog/errata.csp?isbn=9781680452051 for release details.

978-1-680-45209-9

Safari® Books Online

Safari Books Online is an on-demand digital library that delivers expert content in both book and video form from the world's leading authors in technology and business. Technology professionals, software developers, web designers, and business and creative professionals use Safari Books Online as their primary resource for research, problem solving, learning, and certification training. Safari Books Online offers a range of plans and pricing for enterprise, government, education, and individuals. Members have access to thousands of books, training videos, and prepublication manuscripts in one fully searchable database from publishers like O'Reilly Media, Prentice Hall Professional, Addison-Wesley Professional, Microsoft Press, Sams, Que, Peachpit Press, Focal Press, Cisco Press, John Wiley & Sons, Syngress, Morgan Kaufmann, IBM Redbooks, Packt, Adobe Press, FT Press, Apress, Manning, New Riders, McGraw-Hill, Jones & Bartlett, Course Technology, and hundreds more. For more information about Safari Books Online, please visit us online.

How to Contact Us

Please address comments and questions concerning this book to the publisher:

Maker Media, Inc.
1160 Battery Street East, Suite 125
San Francisco, CA 94111
877-306-6253 (in the United States or Canada)
707-639-1355 (international or local)

Maker Media unites, inspires, informs, and entertains a growing community of resourceful people who undertake amazing projects in their backyards, basements, and garages. Maker Media celebrates your right to tweak, hack, and bend any Technology to your will. The Maker Media audience continues to be a growing culture and community that believes in bettering ourselves, our environment, our educational system—our entire world. This is much more than an audience, it's a worldwide movement that Maker Media is leading. We call it the Maker Movement.

For more information about Maker Media, visit us online:

Make: and Makezine.com: makezine.com
Maker Faire: makerfaire.com
Maker Shed: makershed.com

To comment or ask technical questions about this book, send email to bookquestions@oreilly.com.

Please visit http://craftsforlearning.com for more information about the book, and the author.

To my boys: good eaters, good cooks, and always happy to experiment in the kitchen!

Contents

Acknowledgments   x

Preface   xi

Introduction: How to Cook Up a New Invention   xiii

The Edible Inventions Pantry   xvii

1 CRAZY KITCHEN GADGETS AND USEFUL UTENSILS   1

Make a Better Butter Maker   3

Project: Handmade Butter in a Jar   4

Project: Make a Motorized ButterBot   8

3D Food Printing   14

Computers, Drawings, and Pancakes   16

Project: Plot a Drawing on a Graph   17

Project: Make a Hydraulic LEGO 3D Food Printer   20

More about Cooking Tools   36

2 CREATE CHEMICAL CUISINE   37

Gels, Bouncy Spheres, and Rubbery Noodles   39

Project: Juicy Gelatin Dots   41

Project: Agar Noodles   45

Recipe: Agar-Agar Raindrop Cake   50

Foamy Goodness   52

Recipe: Baked Foam Meringue Cookies   53

Recipe: Homemade Whipped Gelatin Marshmallows   58

Recipe: Marshmallow SunPies   61

Recipe: Fizzy Watermelon Lemonade   63

The Chemistry of Crystals   66

Project: Rock Candy Sticks   67

Flash-Frozen Delights   70

Recipe: Dry Ice Sorbet   72

Project: Make Below-Freezing Ice Sorbet   75

More about Chemical Cuisine   77

3   HACK IT FROM SCRATCH   79

The History of Cold Breakfast   82

Recipe: Crunchy Granola   83

The Invention That Changed Baking   85

Recipe: Pancake Mix   86

Recipe: Pancake Mix Pancakes   87

Condiments and Spreads   89

Recipe: Tangy Fermented Ketchup from Scratch   91

Recipe: Oven-Baked French Fries   93

Recipe: Sweet Refrigerator Pickles   95

Recipe: Homemade Hummus   97

Nuke It! The Microwave Revolution   100

Recipe: Applesauce Cake in a Mug   101

Recipe: Homemade Applesauce   103

More about Cooking from Scratch   105

4   GROW YOUR OWN INGREDIENTS   107

Project: Grow Sprouts in a Jar   109

Project: Grow Instant Indoor Veggies from Kitchen Cuttings   113

Project: Plant an (Almost) Instant Avocado Tree   118

Recipe: Guacamole Dip with Tomato   120

Project: Make an Aquaponic Jar   121

Project: Use a Compost Jar to Turn Veggie Scraps into Soil   128

Project: Build an Indoor Worm Bin   129

More about Growing Edible Plants   132

5 COOK OFF THE GRID 133

Cooking with the Sun 136

Project: Build a Cardboard Solar Oven 138

Solar Recipes 146

Recipe: Solar Nachos 147

Recipe: Solar Oatmeal Cookies 148

Recipe: Solar Chocolate Cake 149

Self-Contained Cooking 151

Project: Self-Contained Countertop Yogurt-Maker 152

Project: Put Together a Meal-Sized Thermal Cooker 155

Recipe: Thermal Cooker Lentils and Rice 158

DIY Outdoor Stove 160

Project: Tin Can Cooker 161

Recipe: Grilled Cheese 166

Recipe: Egg-in-a-Hole 167

More about Alternative Cooking Methods 169

Index 171

Acknowledgments

Thanks to the following people for their advice, help, and support in creating *Edible Inventions*:

* Miguel Valenzuela for introducing me to the idea of edible inventions with his original LEGO PancakeBot and designing a scaled-back version for this book that anyone can build—and for suggesting I write this book in the first place

* John Ceceri III for lending his LEGOs and his expertise

* Lily Born and her father Joe for sharing the inspiring story of the Kangaroo Cup

* Amy Halloran for her tips on pancakes and baking soda

* Howard Stoner and David Borton for their tips on solar ovens and tin can stoves

* Kay Holt and Bastian for testing the solar project

* and everyone at Maker Media and Happenstance Type-O-Rama, especially my hard-working editor Patrick DiJusto.

Preface

In 2012, *Make:* magazine asked me how I came to invent the PancakeBot. I told them that one day for breakfast I was drawing pancakes for my daughters Lily and Maia and reading *Make:* magazine. There was an article (http://makezine.com/2014/02/11/blockheads/) that described using LEGO to build a prototype of a machine that stamps images onto pancakes. I don't know exactly how it happened, but I think I mentioned the article to Lily and the next thing I heard was, "Maia! Papa is going to make a pancake machine!" Maia screamed in excitement and both of them did some kind of dance, and the next thing I knew, I was on the hook to make a PancakeBot.

Over the next few years, I brought different versions of the PancakeBot to Maker Faires in New York, San Francisco, France, the UK, and Norway. One of the highlights of that time was when we were invited to the White House Maker Faire in Washington, D. C. Today, I run a company called PancakeBot LLC. We've partnered with another company named Storebound that helps us make and sell machines that automatically "print" pancakes in any design you create. And I've learned a lot about how an invention goes from an idea to a reality.

The thing with inventing is that most inventors focus on, well, the inventing part. As a result, many people struggle to share their inventions with the world. There are lots of way to "take your invention to market," which is when an idea goes from the drawing board to the store shelf (or even an online store). Before sharing your idea, though, you should ask yourself if you want to protect it, or put it out there for the world to run with.

If you want to protect your idea, you'll have to file a patent application. That lets you claim a "patent pending" status and tell the world you intend to protect your idea. Luckily, the United States Patent Office has lots of information on this. You can visit uspto.gov for more information. Another route that you can consider is licensing, where you share your protected idea with a manufacturer who can take it to market for you. In exchange, you get royalty payments based on how many products you sell. (A good reference is the book *Sell Your Ideas With or Without a Patent* by Stephen M. Key.)

If you want to share your idea with a different type of protection, you can make it "open source" and release it with a Creative Commons license. This allows others to use your idea as long as they don't sell it themselves. You can learn more at creativecommons.org. If you're interested in producing the idea yourself, consider a crowdfunding platform such as KickStarter or IndieGoGo. These websites not only help you raise money for your idea, they also allow you to test out what people think about your idea before you actually manufacture it.

Lastly (and most important in my opinion), you have to learn to tell your story. Show people your passion and tell them how you got to your idea. Tell them what inspired you to solve a problem or bridge a gap between two existing ideas. Sharing is a great way to get people excited and build up a good base of fans who want to be a part of your project. A great way to do this is to display your project at one of the many Maker Faires that take place around the world. You'll meet people that have your same passion for creating and making. You may even find someone to help you along the way. Always remember that your idea is just the beginning of a great journey, and persistence is 95 percent of that journey.

Since I built that first PancakeBot out of LEGO for my daughters, I've explored many aspects of food printing, and I'm going to work to keep pushing the boundaries of the technology in the future. My hope is that this book gets you excited about inventing and creating with food, too. What's so special about food? Well, it's something that all of us have in common. It's a personal experience we all share and a way to create and experiment, with a rather low cost. Food binds us together, whether it's when you sit down for dinner with your family or enjoy a treat at a restaurant. Learning about food, how it works, and how you can invent around it opens up your mind to experimenting and making. Inventing with food may help you figure out how to address larger problems and challenges the world faces, or it may just be a way to make you experience cooking in a fun way. Remember that when inventing with food, you can eat your mistakes!

Keep on Making, Creating, and Inventing!
Miguel Valenzuela
Creator of PancakeBot

Introduction: How to Cook Up a New Invention

Believe it or not, there's a lot of inventing going on in the kitchen. Most kitchens today are full of gadgets that can physically transform ordinary ingredients into something new and interesting. They can be as simple as a wire whisk or as complicated as a microwave oven.

Then there's the chemistry involved in cooking. All food—even "natural" and organic food—contains chemicals. That's not a bad thing. Every substance on Earth is made up of chemicals, and chemical reactions are what give cooked foods their taste, texture, and appearance. When you add chemicals to a dish to make it thicker, gooey-er, or puffier, you turn a bunch of plain ingredients into a mouth-watering meal.

Some of the methods you'll get to try in this book go back thousands of years. Others give you a peek at the future of cooking. You'll design machines that make food preparation easier (or just sillier). You'll experiment with chemicals that give food weird, alien forms. You'll discover how microscopic living organisms can add zing to what you eat. You'll explore growing your own food—no farm needed. You'll learn about ways to cook outdoors, without gas or electricity. And you'll get to try out fascinating tools and techniques with delicious recipes.

Edible Inventions: Cooking Hacks and Yummy Recipes You Can Build, Mix, Bake, and Grow will show you some unusual ways to create a meal, and help you invent some of your own. You'll find out first-hand how creative cooking can be, and how far edible inventions can take you.

STEP BY STEP:
THE INVENTION PROCESS

The best thing about inventing is that anyone can do it! But it helps if you know the steps to follow. When Lily Born of Chicago was eight years old, she decided to design a new kind of cup for her grandfather. His hands would shake because of Parkinson's Disease and make it hard for him to hold a cup without spilling it. Lily's cup had three handles that served as legs. When you put it down, the legs kept it steady and upright. She called her invention the Kangaroo Cup, because kangaroos use their tails like a third leg to help them balance.

Credit: Imagiroo

Lily's dad Joe, an inventor, offered to help Lily make and sell her idea. They traveled to China, where Lily was born, to find a company to produce high-quality ceramic Kangaroo Cups. The project was so successful that Lily began to win awards. She was asked to give talks about how kids can become inventors. And in 2015, she was even invited to take part in the White House Science Fair. By the time she was 13, Lily had her own company, Imagiroo, and had sold over 11,000 cups.

How did Lily go from shy school kid to business powerhouse? Her website, imagiroo.com, describes six steps in the invention process:

Observe Take a look around for needs to fill or problems to solve. That's what Lily did when she noticed her grandfather was unhappy because he had trouble doing things for himself.

Brainstorm Think up as many ideas as you can that might solve your problem. It doesn't matter how wild they sound—at this point, anything goes! Write them down or make a sketch of each one, or do both. Then go through your ideas and pick one (or a combination of ideas) to work on.

Prototype Make a model to see how your idea looks, works, and feels. Lily made her first prototype by taking a regular cup and adding three legs using moldable plastic.

Experiment Find volunteers to test your prototype in real-world situations and tell you what they think. Lily used family members as her test subjects.

Iterate Take your testers' feedback and use it to create new versions. You may need to go through the process several times to get a product just right. Lily spent years making clay prototypes at a local pottery studio. "We had enough iterations to fill up five shelves," says Lily. When customers asked for a child-proof model, she went through the entire process again to create a plastic version.

Launch When your design is ready, it's time to do some marketing. Choose a catchy name, design nice packaging, and look for ways to introduce your idea to the public. Lily's dad helped her create crowdfunding campaigns using the websites Kickstarter and Indiegogo. They spread the word about the Kangaroo Cup and the girl who invented it.

You'll find that the steps above are also helpful when trying out the projects in this book. Here are some more tips to keep in mind:

* Keep notes on what you do! Many of the recipes in this book list suggested ingredients, with the amounts left up to you. Write down which ingredients you add and how much. That will help you remember which versions you like so you can make them again.

* A good way to record what you do and how you did it is to take photos and videos. They also make it easy to share your techniques with others or even write your own how-to.

* If a recipe or design doesn't work, don't give up! Go back over the directions and your notes, and figure out some possible solutions to your problem. If you need help, check out books on the subject, search online, or look for cooking experts in your family or community who are willing to look over your work and make suggestions. As Lily says, "Never be afraid to ask for help. Whether you're a kid or an adult, whether it's starting a business or an art project, you're going to need help."

It doesn't matter whether or not you want to start your own company or just learn some new ways to cook. If you like to play with your food, this book is for you. So tie on your apron, pop your chef's hat on your head, and keep reading. *Bon appétit*! (That's French for "Enjoy your meal!")

The Edible Inventions Pantry

Cooking is a lot easier when you have a well-stocked pantry. You probably already have things like salt, pepper, flour, eggs, milk, lettuce, and tomatoes in your refrigerator or on a shelf. If not, they are easy to find in any grocery store. This list includes the less-common tools and ingredients you'll need to make the projects and recipes in this book and where to look for them. You don't need them all for every project—and in many cases, if you have something that will do the same job, it's fine to substitute. In fact, that's one way to invent a new version!

KiTCHEN STAPLES

If you don't already have these items on hand, they're easily found in most supermarkets, discount department stores, and dollar stores. You're sure to find many uses for them in this book and on your own.

INGREDIENTS

* Heavy cream (without additives)
* Liquid food coloring
* Graham crackers
* Powdered ginger
* MolassesVegetable oil, such as corn or canola
* Unflavored gelatin (in the baking aisle)
* Powdered agar-agar (vegetarian gelatin; available at health food stores)
* Cream of tartar (in the spice aisle)
* Lemons or lemon juice
* Powdered sugar
* Baking powder
* Baking soda
* Seeds for sprouting
* White or apple cider vinegar
* Tomato paste
* Pickling salt
* Kosher or rock salt
* Plain or vanilla yogurt with live cultures (preferably not Greek) or probiotic pills
* Chocolate chips

COOKING EQUIPMENT

* Lightweight plastic cup with tight lid
* Empty plastic water bottles (with rings of ridges around the sides)
* Rubber spatula (bowl scraper)
* Microwave oven
* Canning jars (or sturdy glass jars with tight lids)
* Canning funnel (helpful, not required)
* Wax paper or parchment paper
* Food-grade squeeze bottle (such as a ketchup bottle)
* Small to medium wire strainer
* Wine glass with a smooth inner surface (plastic is OK)
* Electric mixer
* Metal spatula
* Ziptop bags, small and large
* Cheesecloth (open-weave cotton fabric—in the cooking or sewing aisle)
* *Optional*: Salad spinner
* Plastic, foam, or paper plates, cups, spoons, and forks
* Skinny or regular drinking straws
* Extra-wide drinking straws
* Oven-proof plastic roasting bag (turkey size is best)
* Aluminum foil
* Hot pads, oven mitts, or dish towels
* Small, lightweight black or dark-colored pans or oven trays (preferably with covers—or make covers by using a second one as a lid, clamped with binder clips)
* Food thermometer with probe (make sure it covers the range of oven temperatures; meat thermometers don't go high enough)
* Quart- (liter-) sized insulated mug or container
* Candy thermometer
* *Optional*: Insulated lunchbag (large enough to hold insulated mug)

CRAFTS / BUILDING / HOUSEHOLD MATERIALS

* Handheld minifan or 1.5 volt DC motor with wires

* AA batteries

* Electrical tape

* Small rubber bands

* Graph paper

* Craft sticks or wooden coffee stirrers

* Heavy winter gloves

* Poster board

* Large plastic bin with top

* Jumbo (half gallon [2 L] or more) glass jars (such as decorative canning jars) or recycled plastic storage jars (half gallon or more)

* White glue

* Masking tape

* Recycled cardboard boxes and scrap corrugated cardboard

* Black construction paper

* *Optional:* Glass from picture frame

* *Optional:* ProtractorBrass paper brad and string

* *Optional:* Peel-and-stick Velcro (hook and loop) tape

* *Optional:* Outdoor rolling cart

* Large, recycled aluminum can

* Tuna-sized aluminum cans

* Large basket or box

* Old, clean towels

* Matches or lighter (use with adult supervision only!)

SPECIALTY EQUIPMENT AND MATERIALS

These are pretty much required for some projects, unless a substitute is mentioned in the instructions. If you can't find them locally, try Amazon.com or other online retailers.

* LEGO Classic Medium Creative Brick Box #10696

* LEGO Baseplate *#10700* (10 inch [25 cm] square, 32×32 studs)

* 4–7 feet (1–2 m) vinyl tubing (¼ inch [6mm] outside diameter—available in the plumbing supply section of hardware stores)

* 2 feet (60 cm) silicone tubing (⅛ inch [3.2mm] inside diameter, ¼ inch [6.4 mm] outside diameter)—look for food grade or peristaltic pump hose online (such as mcmaster.com #3038K13) or at medical supply stores

* 4 10-mL oral syringe with rubber O-rings—some pharmacies will give you free samples

* Large flavor injector food or oral syringe (30 ml or more)—you don't need the needle, if there is one; if you can't find it in the cooking tool aisle, try a pet shop, which carries them for squirting medicine into the mouths of large dogs

* Dry ice—some food or beverage stores, or welding shops

* Fish (zebra danios, small goldfish, or other cold-water breeds)

* Aquarium plant

* Aquarium gravel or pea-sized gardening gravel (rinse well)

* Worms (red wigglers or trout worms)—bait shop, vermiculture retailers, or just under the surface of a pile of leaves or garden trimmings

* Aluminum foil tape (hardware stores in the plumbing and heating aisle)

* Velcro tape (sewing and hardware departments)

* Dark-colored oven pans

* Tin snips

* Heavy safety gloves

* Paraffin wax or old unscented candles

* Double boiler (or two pots that fit one inside the other) to be used for crafts only

* Old frying pan or camping skillet

MEASUREMENTS

Measurements in this book are given in both U.S. customary units (teaspoons, cups, inches, feet) which are used in the United States, and metric units (milliliters, liters, centimeters, meters), which are used everywhere else. Temperatures are given in both degrees Fahrenheit (°F), and degrees Celsius (°C) When reading the directions for a project or recipe, be sure you are looking at the number that matches your measuring spoons or thermometer.

 A FEW COOKING TIPS

- Always read through the entire recipe or project before you start to make sure you have everything you need.

- If you buy more of a perishable ingredient than you need, put the extra in the freezer for future use.

- Taste as you go along, and keep notes on what you like and what you want to change for next time.

- If you're looking for interesting new spice combinations, try using flavored herbal tea as seasoning.

- Don't get frustrated if something doesn't work the first time. Keep experimenting!

A SAFETY NOTE FOR ADULTS TO SHARE WITH KIDS

If you're reading this book, your kids want to cook. Let them do as much for themselves as possible! Keep an eye on them and stay nearby, but put them in charge of the project and give them the chance to try out new skills under your supervision. It helps if you introduce them to some safety rules, and show them the best way to use the stove, oven, microwave, knives, and other kitchen gadgets. Here are some tips to share with your kids:

* Always ask an adult for permission or help before using anything sharp or hot. That includes electric mixers and solar ovens.

* Wash your hands with soap and water before handling food.

* If you need help reaching the counter, make sure you are standing on something sturdy, such as a step stool.

* Keep long hair tied back.

* Tuck in, roll up, or remove anything hanging loose, such as long sleeves, jewelry, or the laces from the hood of a jacket.

* Wear old clothing or cover up with an apron or old t-shirt.

* Use oven mitts or pot holders to grab hot pans, but be sure not to leave them near a hot burner.

* When substituting materials in a project that gets very hot, be sure to use only heat-proof materials.

* To cook in a microwave oven, only use sturdy glass, plastic, or ceramic dishes—flimsy plastic can melt if it gets too hot. Never use metal.

* If you're chopping vegetables, put the item you're cutting on a cutting board and keep the sharp edge of the knife pointed down—making sure there are no fingers in the way! (Many fruits and vegetables can be cut with a rounded dinner knife.)

* Wipe up spills as they happen.

* Be careful plugging and unplugging electrical appliances, and don't use them near the sink.

* Before you leave the kitchen, make sure everything is clean and put away, all food containers are closed up, and all appliances are turned off.

Crazy Kitchen Gadgets and Useful Utensils

Discover what it takes to come up with the next must-have cooking gadget or wacky food machine, and find out how to invent your own!

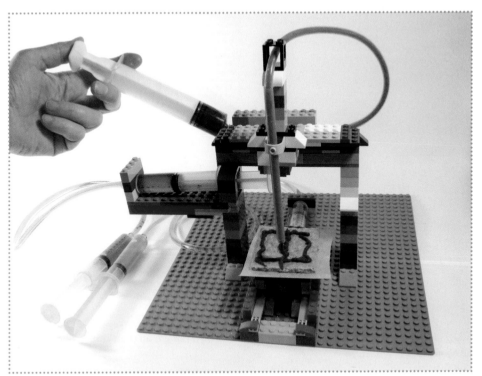

FIGURE 1-1: **A 3D food printer made out of LEGOs? Why not?**

The flood of kitchen gadgets you see advertised every day on TV info-mercials and in trendy housewares catalogs is nothing new. The Industrial Revolution and the rise of factories around 1800 started a boom in "labor saving" devices. Some of these newfangled devices, such as apple corers, egg beaters, and potato mashers, were useful enough to last for many decades. Others were quickly forgotten. They include such strange inventions as glass knives to cut tomatoes (the glass didn't react with the acid in the vegetable and change the color or the flavor, the way metal knives did before stainless steel came along). Or a green bean slicer that required you to insert each green bean one at a time.

It's hard to predict whether today's new inventions—like the combination scissors/spatula for cutting and lifting a slice of pizza in one step, or the chork (half chopsticks, half fork)—will stick around or fade away. To be successful, a new kitchen invention has to improve upon what already exists. It might make food tastier, easier to prepare, easier to handle, or easier to digest. Or it could become popular just by making mealtime more fun. Here are some kitchen gadgets you can make yourself that do a little of both!

FIGURE 1-2: **Behold the chork—perfect for eaters who can't make up their minds between using chopsticks or a fork.**

MAKE A BETTER BUTTER MAKER

The secret of making butter, which is just the fatty part of milk, was discovered early in human history. Chunks of "bog butter" up to 5,000 years old and weighing 100 pounds (45 kg) have been excavated from marshy areas of Ireland, where butter was so valuable it was used to pay taxes. In desert regions, ancient nomadic people carried pouches of milk on their horses or camels and let the motion slosh the liquid around until the fat clumped together on its own. In North America, dairy farmers tra-

FIGURE 1-3: **Fresh local cream separates into cool buttermilk and fragrant butter.**

ditionally used a butter churn: a tall narrow barrel or jar. They filled it with cream (which has more fat in it than regular milk) and then stomped a long wooden stick, called a *dasher*, up and down through the hole in the lid until the butter separated out.

Still, making your own butter took a lot of time and muscle. By the early 1900s, inventors had filed more than 2,500 patents for different kinds of butter-making machines. One popular model was a hand-cranked butter churn invented by Nathan Dazey in 1906. It used wooden or metal blades to stir the milk, like a modern-day mixer. Eventually factory-made butter, easily available on store shelves, replaced the home-made variety.

But do-it-yourself butter is coming back into style. You can whip up a batch in no time with an electric mixer or blender. However, you don't really need any special equipment to make your own butter. Try the following recipe, and then think up ways to design a better butter churn yourself!

PROJECT: HANDMADE BUTTER iN A JAR

FIGURE 1-4: **Real butter is great all on its own.**

Take some cream and shake it up for awhile, and the fat globules will stick together and foam up with air bubbles. (This is how you make whipped cream.) Keep mixing and the fat globules suddenly burst and clump together. Shazam! You've got butter.

The thin watery liquid left over is known as buttermilk. Drinking it is a treat—it has a smooth, buttery flavor, even though the fat has been removed. It's different than the stuff that passes for buttermilk that you buy in a store, which is just low-fat milk with acid added to it to make it thick and tangy. Traditional buttermilk (or the cream it was made from) was left to sit for a while to allow it to ferment, which means allowing friendly bacteria cultures to grow and produce acid naturally. (The same process is used to turn milk into yogurt—see Chapter 5 for a recipe.) Store-bought buttermilk is often used to make pancakes or biscuits because the acid combines with other ingredients to make baked goods rise. (See Chapters 2 and 3 for more on cooking chemistry.)

For best results when making your own butter, look for cream without any added ingredients. Additives like carrageenan prevent the fat from separating out, but now we *want* the fat to separate! Locally produced or organic cream may also have a higher fat content, which will also make your butter extra-rich.

1. Pour the cream into the jar and close the lid. If the lid isn't completely spill-proof, cover the opening first with a piece of plastic wrap and then screw on the lid. Let the jar sit out until it the cream is room temperature. (You can speed things up by placing the jar in a bowl of warm water.)

2. Now shake the jar! That's it. The shaking process can take anywhere from five minutes to half an hour. It depends on the cream, the container, and how strong your arm is (or how many friends you have to help you). As you shake, you're mixing in air that makes the cream thick. (Add sugar at this point and you have whipped cream!)

3. Keep shaking until you start to hear and feel something thunking against the inside of the jar. Take a peek inside. If you see a large yellowish ball floating in some pale liquid, the fat has separated out from the buttermilk.

4. Pour the buttermilk into a clean glass to drink or use in baking. To help the butter stay fresh longer, remove as much of the remaining buttermilk as you can. Fill a bowl with ice-cold water and dunk the butter

FIGURE 1-5: **The handle on this cup makes it easier to shake.**

FIGURE 1-6: **The butter has "seized" and formed itself into a yellow clump.**

ball. The cold water will make the butter harden quickly and rinse away any buttermilk still clinging to the ball. Pour out the cloudy water and repeat once or twice until the water runs mostly clear.

5. Store the butter in an airtight container. It will keep it in the refrigerator for a couple weeks.

6. *Extension*: To make flavored compound butter, trying adding some of the following ingredients while the butter is still warm and soft:

- Basil
- Garlic
- Chives
- Honey
- Maple syrup
- Cinnamon
- Strawberry jam
- Orange marmalade
- Sweetened cocoa

FIGURE 1-7: **Dunk the butter in water to rinse off the buttermilk.**

FIGURE 1-8: **Mix in honey or other flavorings to make compound butter.**

SUPER SPEEDY BUTTER IN A BOTTLE

The traditional way of making butter in a glass jar can take 20 minutes of shaking or more. Using a lightweight plastic jar can cut that time in half (perhaps because it's lighter and your arm doesn't get tired as quickly). But a simple trick can shrink shaking time to just a minute or two. All you need is an empty water bottle (the kind with ridges around the middle), a sharp knife, and, if you're a kid, an adult to help with the knife. Pour the cream into the bottle and follow the directions for the Butter in a Jar project. The ridges on the water bottle help break up the fat globules faster than the straight, smooth sides of a jar. When the buttermilk has separated from the fat, turn the bottle over and let the thin liquid pour out the spout. Then, use the knife to cut off the bottom of the bottle so you can get the ball of butter out. It's that easy!

FIGURE 1-9: Look for a water bottle with ridges.

FIGURE 1-10: Be careful cutting the butter out of the bottle.

PROJECT: MAKE A MOTORIZED BUTTERBOT

🥄 MATERIALS

2 sturdy, lightweight, stackable cups with tight-fitting lids, such as recycled drink cups or food containers

3–4 thick rubber bands that fit snugly around cup

3–8 craft sticks

Optional: Googly eyes, pipe cleaners, etc.

Dollar-store handheld mini-fan or small DC motor (repurposed from an electric toothbrush or old toy is good—attach wires if needed)

1–2 AA batteries (depending on what your motor requires) plus extras

Optional: Small rubber band

Optional: Foam tape

Electrical tape

Medium cork or piece of non-drying clay

Optional: Large wooden beads, hot glue gun

Lightweight plastic fork

¼ cup (75 ml) heavy cream

Optional: Plastic wrap (to cover the opening of the cup with before snapping on the lid if it is not completely spill-proof)

If you had to design a new type of butter churn, what method would you use? Would you focus on creating a more efficient dasher? Or try to find a completely different way to shake up the cream—like turning the butter-maker into a ball you can toss around or an RC vehicle you can roll back and forth? This Motorized ButterBot may not be the most efficient butter-maker ever built, but it definitely has the most personality. It's begging to be fancied up with whatever you have on hand, so use your imagination! Remember, if your ButterBot doesn't work the first time, keep trying—it's all part of the invention process.

FIGURE 1-11: **A ButterBot made with a DC motor and battery**

❧⟶ BUTTERBOT TO ARTBOT

This ButterBot is a variation on the Vibrating ArtBot project in *Robotics: Discover the Science and Technology of the Future* (Nomad Press, 2012). If you replace the craft stick legs with skinny markers, you can make it draw while it churns your butter!

1. Set aside one of the cups and its lid to use as a washable insert. Take the other cup and slip the rubber bands around it. Slide craft sticks under the rubber bands so they are evenly spaced around the cup. Adjust them so they hold the cup up, like legs. Add any other decorations you like, such as googly eyes or pipe-cleaner arms.

2. If using a mini-fan, insert the required batteries. Remove the fan blades from the motor shaft (the thin metal rod that spins).

3. Attach an off-center weight to the shaft of the motor. If using a cork, just push it onto the shaft so that more hangs off one end than the other. You can hot-glue a craft stick or beads onto the cork for added weight. If using a piece of clay, stick it on and shape it so it is heavier on one side. Be careful not to jam up the motor with the clay or hot glue, or the shaft will have trouble spinning.

4. Attach the fan to the lid of the insert cup. Make sure it is secure enough to stay on when the ButterBot is shaking. You can use double-sided foam tape as well as electrical tape, but keep in mind that you may need to open up the fan to change the batteries before your butter is done. Check that the weight on the shaft can spin freely without

FIGURE 1-12: **The legs on the ButterBot raise the center of gravity so it shakes more than on the cup's stable bottom.**

FIGURE 1-13: **You can pull the fan's blades right off, along with the center piece that holds them in place.**

hitting the cup or causing the cup to tip over. You also need to be able to open and close the lid easily.

5. If you are using a DC motor, first attach wires if it doesn't already have them. Take two short, insulated wires and twist the metal ends onto the positive and negative terminals of the motor. Secure them with electrical tape. Test your motor to see if it can run with one battery by touching the ends of the wires to the ends of one of the batteries. If not, try connecting two batteries pointing in the same direction—use a craft stick as a "splint" to keep them straight and wrap them mummy-style with electrical tape. Loop a thick rubber band around the ends of the battery (or around your double battery pack, if needed), and secure with more electrical tape. Slip one of the motor wires under the rubber band of your "battery pack" so the metal wire touches one of the metal battery terminals, and tape it in place. Slip the other end under the rubber band by the other terminal of the battery. This is your On/Off switch; to stop the

FIGURE 1-14: **Make sure whatever weight you add to the motor shaft is off center—that's what causes it to shake.**

FIGURE 1-15: **If you're using a fan as a motor, attach it well with plenty of tape.**

FIGURE 1-16: **The wires on this motor were curled up by wrapping them around a pencil to take up any slack.**

motor, just disconnect the wire from the battery. Follow the directions in Step 4 to attach the motor and batteries to the cup lid.

FIGURE 1-17: **The fork keeps the cream from just swirling around.**

6. You're ready to make some butter in your ButterBot! Place the insert into the decorated cup. Then take the plastic fork and carefully cut or break off the handle so that it will fit inside the cup with the lid closed. Place the fork into the cup, pointy end down, to help stir the cream as it shakes. Pour in the cream and close the lid securely. If the lid isn't absolutely spill-proof, cover the opening first with a small piece of plastic wrap before closing the lid.

7. To keep the ButterBot from traveling around the table or falling over as it shakes, put it in a small box or other flat container. Then turn on the motor. It may take an hour or more for the ButterBot to churn the cream into butter. Keep it within sight while it is turned on, and check it often to make sure it is working correctly. If the motor or battery gets too hot, turn the ButterBot off and let it cool down for a few minutes. You may need to pour out some of the cream to lighten the load.

8. After a while, you will notice that the cream has gotten so thick it doesn't slosh around anymore. Watch the Butter-Bot carefully from this point on. When the butter "seizes," it will suddenly separate into a ball of butter and thin buttermilk. This may throw the ButterBot off balance and cause it to tip over!

FIGURE 1-18: **The craft-stick weight on this ButterBot swings around.**

FIGURE 1-19: **It takes longer, but the cream will separate.**

9. When the butter is finished, rinse and store it, following the previous directions for making butter.

3D FOOD PRINTING

It sounds like science fiction, but the food gadget of the future may be a 3D food printer that does all the cooking for you. Regular 3D printers work a little like document printers, but instead of ink, they squirt out a thin stream of softened plastic. The plastic builds up layer by layer to create models of 3D objects like toys, phone cases, and robot parts. Experimental 3D food printers make edible designs out of chocolate or sugar that would be impossible using traditional candy molds. Someday soon, you may be able to insert a capsule filled with different ingredients into a 3D food printer to produce pizza, cookies, or other snacks. In fact, scientists at the United States space agency NASA are looking at using 3D food printers to produce a variety of meals for astronauts on long missions.

FIGURE 1-20: **PancakeBot inventor Miguel Valenzuela with his daughters Lily and Maia** *Credit: Miguel Valenzuela*

But if you just can't wait, there's one 3D food printer you can use right now at home: the PancakeBot. It can copy a computer image using pancake batter and cook it up on a heated griddle, ready for eating. The PancakeBot was invented by engineer Miguel Valenzuela of Norway, who got the idea from his daughters Lily and Maia when they were small. (See the preface for the story of how it was invented.)

Valenzuela wanted to design something that other people could build at home too. So the original PancakeBot was "99 percent LEGO, 1 percent ketchup bottle." The frame was made of LEGO bricks, and its brain and motor came from a LEGO Mindstorms robotics kit. The PancakeBot was the hit of World Maker Faire New York in 2012, but sadly, the LEGO structure didn't survive the trip back to Norway. Valenzuela built a second version out of sheets of clear acrylic plastic. Then, in 2015, he launched a Kickstarter crowdfunding campaign to create a factory-built PancakeBot you could buy to use at home. The campaign raised nearly half a million dollars, and the PancakeBot spread to kitchens around the world.

FIGURE 1-21: **The version of the PancakeBot you can buy for your home**
Credit: PancakeBot LLC

COMPUTERS, DRAWINGS, AND PANCAKES

The trick to making pancake art is to fill a squeeze bottle with pancake batter and draw the lines first. Then you can go back and pour more batter into the open spaces between the lines. The lines get darker because they have cooked longer, so you end up with two-toned drawings. (Try it yourself using the pancake mix recipe in Chapter 3.)

Of course, it's a little more complicated to get a machine to copy a drawing instead of doing it yourself freehand. The PancakeBot has a computerized brain that tells it where to move its nozzle and when to squeeze the batter out. So the drawing must be translated into code the computer can understand. For line drawings, the computer breaks the line down into a series of dots or points. Each point has a set of *coordinates*—numbers that tell the computer where the point is located. It's a lot like number lines and graphs in math. To see how it works, translate a simple drawing into coordinates using pencil and paper.

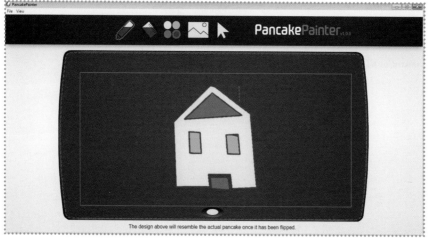

FIGURE 1-22: Pancake Painter is a drawing program you can download to your computer, even if you don't own a PancakeBot.

PROJECT:
PLOT A DRAWING ON A GRAPH

Graph paper is divided into boxes by a grid of *horizontal* (side-to-side) and *vertical* (top-to-bottom) lines. You can use this grid of lines to simulate the way a computer translates a drawing into a bunch of points, each with its own coordinates. (If you don't have graph paper handy, take regular lined paper and use a ruler to draw evenly spaced lines that go up and down across the printed lines.)

MATERIALS

Graph paper

Optional: Lightweight plain paper (thin enough to see the graph paper through it) and paper clips or tape

Ruler

Pencil

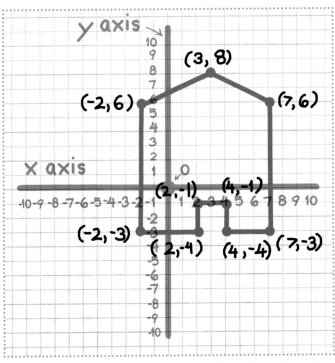

FIGURE 1-23: **Coordinates are sets of numbers that tell you where a point is on a grid.**

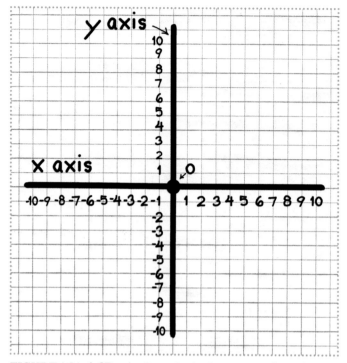

FIGURE 1-24: **A 2D pair of axes is also known as a Cartesian plane.**

1. Use the pencil to make a dot in the middle of the paper where two lines cross. Trace over the lines that cross at that spot. The horizontal line is the X axis. The vertical line is the Y axis. (If this were a 3D drawing, a piece of paper lying flat on a table would have a third line pointing toward the ceiling and the floor—the Z axis.)

2. Next, start to write numbers on the X axis. On the dot in the middle, write a small number 0. Then start to move to the right and number the lines that cross the X axis 1, 2, 3, and so on. When you reach the end, go back to 0. Begin to move to the left and number the lines the same way, but make each

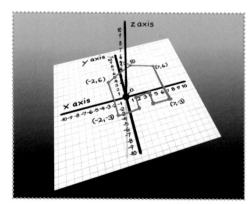

FIGURE 1-25: **The Z axis lets you show points that are above or below the X-Y plane.**

number negative. In other words, you will write -1, -2, -3, and so on.

3. Do the same with the Y axis. The numbers going up from 0 are positive (1, 2, 3) and the numbers going down are negative (-1, -2, -3).

4. If you want to reuse your graph for other drawings, place the plain paper over the graph paper. Use paper clips or tape to keep the paper in place.

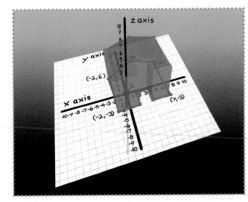

FIGURE 1-26: **How would you write the coordinates of a point on top of the 3D purple house? Hint: It should be in the form (X, Y, Z).**

5. On the plain paper or graph paper, make a simple drawing, such as a smiley face or a house. Now make dots on your drawing wherever it crosses over the intersection of two lines on the graph paper below. Then, plot the points by writing down the numbers that show where each point is. Coordinates are written in the form (X, Y), with the X coordinate always coming first. For instance, if your drawing overlaps the place where the 2 line on the X axis intersects the 5 line on the Y axis, the coordinates of that point are (2, 5).

6. *Extension*: The PancakeBot uses a computer-aided design (CAD) program called Pancake Painter. Try playing around with it, even if you don't have a PancakeBot. You can download the Pancake Painter software for free at http://www.pancakebot.com/download-software.

⇒⭑ LEGO GRAPHiNG

You can label coordinates on a LEGO baseplate the same way you do on graph paper. Instead of counting lines, however, you count the studs. If the bottom edge of the plate is the X axis and the left edge is the Y axis, you can number each row of studs with positive numbers: 1, 2, 3, and so on. You can also find the coordinates along the Z axis by counting each brick you stack on the baseplate as one unit.

PROJECT: MAKE A HYDRAULIC LEGO 3D FOOD PRINTER

MATERIALS

LEGO bricks:

 LEGO Classic Medium Creative Brick Box #10696

 LEGO Baseplate #10700 (10 inches (25 cm) square, 32× 32 studs)

 Optional: LEGO 10×10 octagonal plate #6037610

4–7 feet (1 m) vinyl tubing (¼ inch [6 mm] outside diameter—available in the plumbing supply section of hardware stores)

Optional: **3 feet silicone tubing to use for the food tube instead of vinyl tubing (¼ inch [6 mm] outside diameter, ⅛ inch [3 mm] inside diameter, such as mcmaster.com #3038K13)**

4 10-ml oral syringe (ask your pharmacist for free samples or order online)

30-ml food or oral syringe (look in kitchen supplies at the dollar store, or at a pet shop)

Edible drawing surface, such as a piece of graham cracker, a brownie, or other flat, dry food

Edible "ink," such as cake frosting, chocolate hazelnut spread, or other soft, thick, edible paste

Wax paper

Extra-wide drinking straw (big enough for food tube to fit inside)

This LEGO model of a 3D food printer can be used to make designs with soft, edible material, such as frosting, on an edible surface, such as a graham cracker. Its basic design was created by Miguel Valenzuela, the inventor of the PancakeBot. Although it doesn't have a motor or a computer brain like the real PancakeBot, it will help you explore how 3D printers use three-axis coordinates to make their designs. You'll also learn about *extruders*, machine parts that push softened material through a tube or hole to shape it. On a regular 3D printer, the extruder is a heated nozzle that melts plastic thread and squirts it out in layers to make different shapes. Food extruders in factories squeeze soft ingredients through specially shaped tips to make snacks, cereal, and other foods.

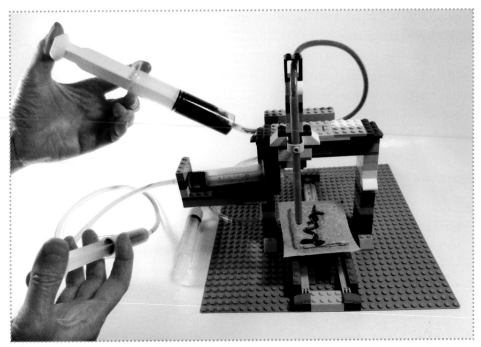

FIGURE 1-27: **With two hands, you can control the flow of the frosting and move the printer food tube back and forth at the same time.**

There are two parts that move on the LEGO 3D food printer. The food tube is the extruder that squirts out the frosting. It is attached to an arm that moves side to side along the X axis on a crossbeam held up by two towers. The print bed holds the graham cracker. It moves back and forth along the Y axis on a set of rails attached to the baseplate. You can copy a simple design like the one you drew on the graph paper by combining these two movements, the same way you draw on an Etch A Sketch. To make the food printer truly 3D, you can also move the food tube arm along the Z axis by pulling the tube up or down through the drinking straw that supports it. This lets you make your printed design taller by piling one layer on another.

Instead of a motor, you move the LEGO 3D food printer parts around by hand, using hydraulic power. *Hydraulic* systems work by pushing a fluid back and forth through a tube. In factories and auto repair shops, they are used to move and lift heavy weights. The LEGO 3D food printer has two hydraulic systems, one for the food tube and one for the print bed. Each is made up of a piece of plastic tubing with a syringe on both ends. When you fill the system with water and push in the plunger on one syringe, the water pushes out the plunger on the other end. One syringe is used as a controller, and other is the piston—the part that moves things around.

The printer also uses a syringe to squeeze the frosting out through the food tube. When you push the plunger, the frosting squirts out of the food tube onto the graham cracker. To move the food tube around using both controllers and push out the frosting at the same time, you would need three hands. You may want to get a friend to help!

TIPS FOR BUILDING WITH LEGO

Here is an explanation of the numbers and terms used to describe LEGO pieces.

- In LEGO terms, a *brick* is a piece that is 1 cm high. A *plate* is flatter than a brick; it takes three plates stacked up to match the height of a brick. A *tile* is like a plate except that the top is smooth. LEGO pieces usually have *studs* (little round bumps with flat tops) and/or *knobs* (short hollow tube-like bumps) that let you connect them to other pieces.

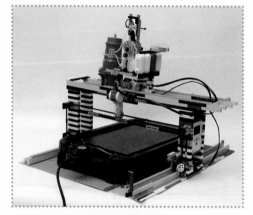

FIGURE 1-28: **The original PancakeBot was built using LEGO bricks and was activated by a LEGO Mindstorms robotics set and a pump to squeeze out the pancake batter.**

- The size of a LEGO piece is described by the number of studs it contains. The shape is assumed to be a rectangle unless another shape is specified. For example, a 2×6 piece is a rectangle containing two rows with six studs. This description is used for bricks, plates, and other pieces that cover the same amount of space.

- Set numbers and element ID numbers are given where needed to make it easier to find or order the parts you need. (The color does not matter.) You can find all the pieces you need to build the LEGO 3D food printer in the LEGO Classic Medium Creative Brick Box set (#10696), except for a 32×32 base-plate (#10700).

- You can also build this model using pieces from your own LEGO collection. Just look at the function of each section of the machine, as described in the directions, and adapt the design to fit the pieces you have on hand.

- If you don't have the right pieces to make holders for the syringes, you can connect them to the model with rubber bands or zip ties.

- If you don't have the pieces to make a slot, build one out of two lines of bricks or plates.

- Wherever possible, try to substitute longer pieces where the sample here shows many shorter pieces put together. This will make your model more stable.

- You can find the set and baseplate in the toy department of many major stores, or order them from shop.lego.com. If you need any extra individual pieces, you may be able to order them from shop.lego.com/en-US/Pick-A-Brick-ByTheme or wwwsecure.us.lego.com/en-gb/service/replacementparts/sale.

- LEGO Digital Designer is a CAD program just for LEGO creations. Although it is no longer supported with updates by LEGO, you can still download it and use it on your computer.

PRINT BED AND RAILS

1. First make the separate print bed. The top is two 4×6 plates is (or one 8×6). It is held up by runners along each long edge that are one stud wide. Each runner is eight studs long and two bricks high, with a 1×6 plate centered on the bottom.

2. Now, put together a group of pieces that will act like a slot for the handle of the syringe plunger. Snap a 1×4 ball socket plate (#4667166; also called a coupling link) into the

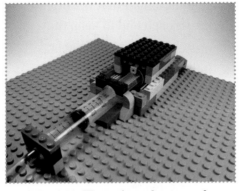

FIGURE 1-29: **The syringe plunger pushes the printer bed back and forth along the rails.**

studs on a 1×4 brick with 4 knobs (#4164073). Then, snap the knobs into the bottom of the print bed at one end. The socket plate should be facing out.

3. Next build the rail set. Starting at one corner, leave a space of nine studs along one edge. Then make two 2×16 rows of bricks going back from the edge with a space of four studs between them. On top, on the inside edge of each rail, lay a row of tiles one stud wide to make a smooth surface for the print bed runners to slide along. It should be at least 12 studs long. You can skip the stud on the outside edge. If you don't have enough tiles, you can use some 1×2 radiator grilles (#241221). Next to the row of tiles, lay a row of one-stud-wide bricks at least 12 studs long. Put a 1×2 brick across each end to keep the print bed from going too far.

4. Between the rails, put a 2×8 row of tiles (made up of two 2×4 tiles) on the baseplate. This lets the syringe plunger slide back and forth smoothly. To stop the plunger from going too far, put a 1×2 plate across each end.

FIGURE 1-30: **The printer bed is a separate piece that sits on the rails.**

FIGURE 1-31: **The brick with knobs snaps onto the socket plate.**

FIGURE 1-32: **The knobs on the top of the yellow plate let you snap it onto the underside of the print bed.**

Then, put a 1×1 brick or 1×2 45°
roof tile (#6022005) on each
side between the plate and
the rail.

5. Build a holder to keep the
front of the syringe in place.
Right up against the end of the
set of rails, snap two curved
2×4×²⁄₃ plates with bow pieces
(#4651237) against each other
on the baseplate so they form
a V. Around it, build an arch.
On either side of the V, stack a
1×2 roof tile and a 1×1×1 ¹⁄₃ brick
with arch piece (#4655246).
Connect them with a 1×4 brick
with bow (#4623775).

6. Leave a space of six studs
behind the V-shaped pieces,
and build a holder with a dia-
mond-shaped opening to hold
the nozzle of the syringe. The
bottom is two 1×2 roof tiles,
slanted sides touching. The top
is two 1×2 inverted roof tiles
(#366526). Cap them with a
1×4 plate.

FRAME

1. Start by building the towers
that hold up the frame. The
backs of the towers line up
with the back of the print bed
rails. Leave a two-stud space on

FIGURE 1-33: **The tiles make the rails
smooth enough for LEGO pieces to slide on.**

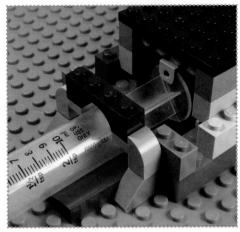

FIGURE 1-34: **The arch and bow pieces hold
the syringe in place.**

FIGURE 1-35: **The inverted roof tiles have
holes on the bottom that let you snap them
onto other pieces.**

the baseplate between the rails and the towers. When they are done, snap a 2×2 roof tile onto the baseplate in front of each tower for stability.

2. For the tower on the right, stack 12 2×4 bricks. If you want, replace one with a slanted or other fancy brick for decoration.

3. The tower on the left supports a ledge that holds the syringe that pushes and pulls the food tube. Stack the following pieces, starting from the bottom, as shown in Figure 1-39:

- Eight 2×4 bricks

- One 2×4 plate

- One 2×3 brick and a 2×3 25° inverted roof tile (#4500469)

- One 2×8 plate

- One 2×2 plate

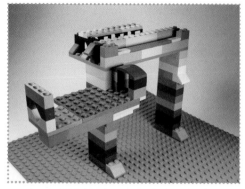

FIGURE 1-36: **The ledge on the left holds another syringe to move the food tube arm.**

FIGURE 1-37: **You can substitute a regular brick for the row with the white roof piece.**

FIGURE 1-38: **The completed ledge**

FIGURE 1-39: **The left tower up to the support**

4. The ledge needs to be 12 studs long and stick out a width of four studs from the front of the tower. Next to the 2×2 plate, put a 6×8 plate so that the long side hangs out to the left of the tower. To make the ledge long enough, snap on a 2×8 brick underneath, so that it can hold a 4×4 plate to the left of the 6×8 plate. Fill in the space between the brick and the tower with additional plates for stability. Then finish the tower with two more 2×4 bricks.

FIGURE 1-40: **The ledge is made of two plates, with supports underneath.**

5. On the ledge, build holders for the syringe. The back holder is the same as the one shown in Step 6. The front holder is an arch. Stack two 1×2 bricks on either side of the ledge so that they hang off the front. Connect them with two plate-with-bow pieces (#4649773). Snap a 1×4 plate underneath for stability.

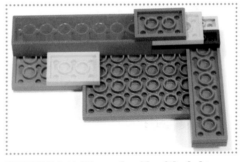

FIGURE 1-41: **The underside of the ledge.**

6. The crossbeam that connects the two towers holds the rails for the food tube arm to slide on, and the guide for the food tube. Start with an 8×16 plate. Along the front edge, make a rail from 1-stud-wide bricks. At each end, snap on a 1×2 brick. In between, top the rail with tiles one stud wide. Leave a

FIGURE 1-42: **A skinny plate under the arch butts up against the ledge. If you have extra pieces, connect it to the ledge from underneath.**

space one stud wide behind the front rail. Build a second rail two bricks high. Connect the two rails with a 3×16 cap made from available plates. Set aside for now.

7. Make the guide for the food tube. In the rear-left corner of the crossbeam plate, facing left, make an arch from two 1×2 roof pieces and a 1×4 brick with bow. Skip a row of studs, and then put together two 2×2 wall element with window pieces (#4539128) so they form a box that lines up with the opening of the arch. Make two more boxes the same way, leaving a one-stud space in between them. Connect them to each other and the arch with a cap made of long plates.

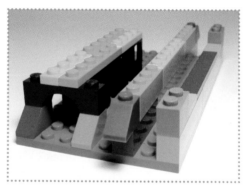

FIGURE 1-43: **The tube guide and the rails for the food tube arm**

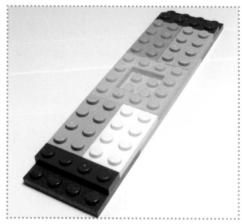

FIGURE 1-44: **The cap is a mishmash of available plates.**

FIGURE 1-45: **Windows are repurposed to hold the food tube in place.**

FOOD TUBE ARM

1. To build the food tube arm, start with two 2×6 plates (or one 4×6 plate) and stack the following pieces on it, starting from the bottom:

   - One 2×4 brick across the front (connecting both plates, if using two)

   - One 2×4 plate.

   - A row with a loop to hold the food tube, consisting of:

     - Two 2×2 roof tiles, hanging off the sides

     - Between them, along the back, two 1×4 bricks with 1 knob, facing forward.

     - One 1×2×2⅔ window frame (#4623514) snapped onto the knobs, with the loop sticking out the front

   - One 2×4 plate

   - Four 2×2 bricks

   - One 2×3 brick with arch (#621501), with the rounded side facing front

   - Two 1×2×2 wall elements with window pieces put together to make a box, like on the food tube guide

FIGURE 1-46: **A rounded brick below the top helps the tube bend down and through the loop (actually another repurposed window frame).**

FIGURE 1-47: **The white slanted tiles are just for decoration.**

2. Make another slot like you did in Step 2 of the Print Bed. Attach it underneath the food tube arm, facing left. For extra security, you can add a row of 1×1 angular bricks (#4187334) behind it, attached to the plates above. Add 1×4 plates in front of and behind the group of pieces for the slot.

3. At the other end of the food tube arm's base you need an overhang and tiles to help it slide along. Underneath the base, along the back edge, center a 1×6 brick. Then put a 1×2 tile on top of the base, also centered. Next, put two 1×4 bricks with bow pieces (#6045945) on either side of the tile, snapping it into the pieces below. They will hang off like two wings.

4. Finally, seat the food tube arm onto the crossbeam. Insert the overhang part of the food tube arm into the space between the rails. Then, attach it firmly to the rails. Make sure the food tube arm can slide smoothly from side to side. If not, check that all the pieces are pressed firmly into place against one another.

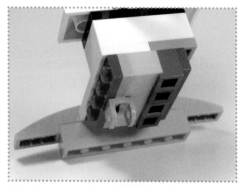

FIGURE 1-48: **The slot has angular bricks that help hold it up and extra plates front and back.**

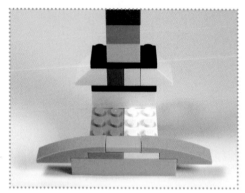

FIGURE 1-49: **The "wings" help balance the food tube arm.**

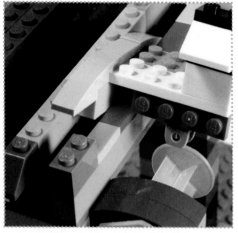

FIGURE 1-50: **The overhang on the food tube arm fits snugly in the opening.**

HYDRAULIC SYSTEMS

1. You need to make two hydraulic systems. Each one consists of a piece of vinyl tubing about 2 feet (60 cm) long with syringes inserted into the ends. Fill each system by holding it over a sink with the syringe plungers pointing up. Push one plunger all the way in. Pull the other plunger out completely. Carefully fill the open syringe with water all the way to the top. (You can put a drop of food coloring in the water to make it easier to see.) Slowly put the plunger back in and push it all the way in. Let the plunger on the second end come all the way out. If there is space in the open syringe, add more water. Then, replace the second plunger. Repeat until you've gotten as much air as possible out of the system. If it is too full, pour out a tiny bit of water. Once it's working, be careful not to push or pull the plungers too far.

2. Put the first hydraulic system into place behind the print bed. The plunger of the syringe in the holder should sit on the strip of tiles on the baseplate between the rails. Seat the print bed on the rails so that the slot is over the top of the

FIGURE 1-51: **The hydraulic system at work**

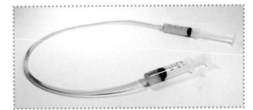

FIGURE 1-52: **Use two different colors to help you tell one system from the other.**

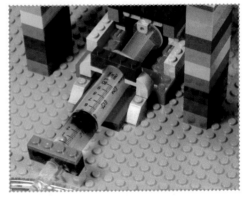

FIGURE 1-53: **A view from inside the rail set shows where the top of the syringe fits.**

plunger handle. Put the second hydraulic system into place on the ledge, with one syringe locked into the holders and the plunger facing toward the food tube arm. Carefully insert the plunger handle into the slot underneath the food tube arm. Make sure to fix any pieces that are loosened.

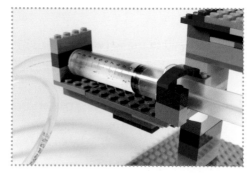

FIGURE 1-54: **The syringe on the ledge**

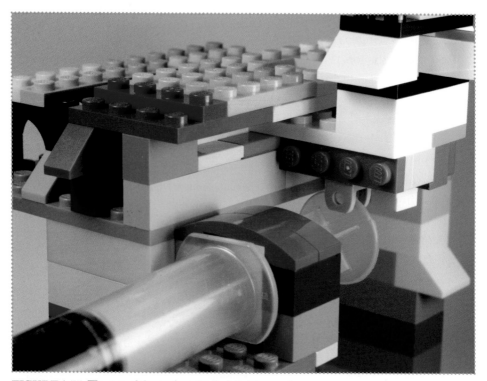

FIGURE 1-55: **The top of the syringe fits behind the arch.**

PRINTING

1. Now it's time to prepare the printing materials. Break some graham crackers into squares, or use other flat, dry food as your edible printing surface. Cut some squares of wax paper to put under the crackers, to protect the LEGO pieces. Place an extra square of wax paper on the print bed to catch drips from the food tube until you put a cracker on it.

FIGURE 1-56: **It takes practice to get the frosting where you want it.**

Tip   You can make a bigger print bed if you have larger plates on hand, such as the 10×10 octagonal plate (#6037610). Snap it right onto the 6×8 plate.

2. Attach the food tube to the large food syringe. Pull out the plunger, but leave the tubing attached. Take some frosting, chocolate hazelnut spread, or other thick food paste. Thin it with a tiny bit of water and stir until it is smooth. It should be squishy, but not runny. Scoop some into the food syringe and replace the plunger. Test it by

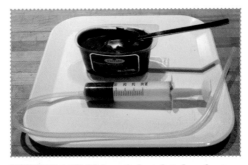

FIGURE 1-57: **Thinned chocolate hazelnut spread works very well.**

squeezing some of the frosting out of the tube onto a plate or graham cracker. If the plunger is hard to push, stir in a little more water until you have a good consistency.

3. Cut the straw so it is 5 inches (13 cm) long. Thread the food tube through the guides on the crossbeam and through the guide on top of the food tube arm. Stick the straw into the loop on the front of the food tube arm,

and insert the tube into it. The straw should stay in place by itself, but if not, use a little tape. Point the food tube straight down. Practice moving the print bed and food tube arm with the loose syringes (which are the controllers) by pushing and pulling on the plungers. It's easiest to start with the controller plungers pulled out as far as they can go. Place them together where you can reach them easily.

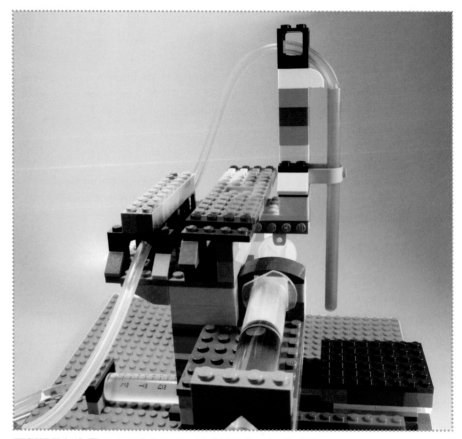

FIGURE 1-58: **The straw is loose and held in place by the tube.**

4. Place a cracker and wax paper onto the print bed. Get a helpful friend (if you have one nearby) to start pushing the food syringe plunger in very slowly, while you work the controllers. If not, take the food syringe in one hand and one of the controllers in the other. Start pushing them both slowly, and draw with the frosting as it comes out. Switch

controllers to move the food tube in the other direction. The more you practice, the better your designs will get.

5. *Extension:* Make a Z-axis controller by adding a third hydraulic system to raise the food tube up and down. Attach it to the food tube arm, and attach the tube directly to the plunger.

MORE ABOUT COOKING TOOLS

Consider the Fork: A History of How We Cook and Eat by Bee Wilson (Basic Books, 2012)

Kangaroo Cups, with free printable Invention Process workbook, invented by Lily Born (imagiroo.com)

PancakeBot (pancakebot.com)

3 Digital Cooks, a network featuring 3D food printer news and resources from around the world (3digitalcooks.com)

Create Chemical Cuisine

2

When cooks start playing around with science, the results can be pretty surprising. Try some of these weird recipes and see for yourself.

FIGURE 2-1: **Cooking with chemistry is fun—and delicious.**

Good cooks and mad scientists have a lot in common. They both like to play around with substances and equipment and see what happens. At fancy restaurants, adventurous chefs are inventing new techniques and whipping up meals no one has ever tasted before. Would you try something called Mustard Air? How about Cucumber Sherbet, Bacon Powder, Grape Sponge, or Chilled Orange Soup? Do Carrot Pillow, Cocoa Caviar, or Butter Foam make your mouth water? They're all part of a gourmet trend known as *modernist cuisine* or *molecular gastronomy*. Its goal is to take familiar flavors and present them in unexpected forms—the more bizarre the better. Thanks to the magic of high-tech ingredients, and equipment like whipping siphons, which shoot gas bubbles into drinks and other foods, modernist cooks are transforming normal ingredients into otherworldly textures and shapes like ices, powders, pearls, noodles, jellies, and foams.

The following recipes let you experiment with modernist cuisine. All you need are a few ingredients you can easily find in your local supermarket or health food store, and tools you probably already have in your kitchen, like an electric mixer and blender. Some of the recipes will be familiar. Others are odd, exciting, and hopefully tasty. Give them a try, and then think about how you can invent some new foods of your own!

SAFETY TIPS FOR WORKING WITH HOT AND STICKY MIXTURES

Young chefs will need adult supervision for all recipes in this chapter, especially those that use electric mixers and blenders, the stove and microwave oven, and super-cold substances. Always use oven mitts or hot pads to handle pots and pans on the stove and to remove containers from the microwave. We recommend using glass containers in the microwave instead of plastic—you don't want your bowl to melt! For recipes that specify a microwave, it's OK to heat liquids on the stove instead—just be sure to stir them constantly to prevent burning. Also, to keep your counters from getting sticky, it helps to put out an extra plate or two to hold gooey spoons and other tools.

GELS, BOUNCY SPHERES, AND RUBBERY NOODLES

Modernist cooks use gels a lot. They're easy to mold into different shapes, they have a clean taste that lets the flavor of the main ingredients come through, and they're fun. In scientific terms, gels are a type of *soft matter*. That means that they don't flow out of a container when you try to pour them like a liquid, but they don't stand stiffly on their own like a solid. Instead, they tend to jiggle. Gels are basically liquid with tiny particles of solid matter spread evenly throughout. As they set, the solid particles link up with each other to form a support structure that helps the gel hold its shape, somewhat.

Powders that create a gel when mixed with liquid are known as *hydrocolloids*. Molecular gastronomy uses many kinds of hydrocolloids. In this chapter, you'll use two of them, *gelatin* and *agar-agar* (usually just called *agar*). Gelatin is made from a fiber-like protein found in meat and bones. You can usually find unflavored gelatin next to the Jell-O in your local supermarket. Agar comes from seaweed, so it's popular with vegetarians. You can find powdered agar in health food stores.

FIGURE 2-2: **Mint tea–flavored gel noodles**

FIGURE 2-3: **What unflavored and uncolored gelatin looks like when you first mix it up**

✎➤ CHEMICAL ENEMIES

Although you can experiment with any kind of flavoring or ingredient in modern cuisine, avoid using pineapple and gelatin together. The pineapple has certain chemical enzymes that will break down the molecular bonds holding the gelatin together and turn it into runny goop. If you're curious, try it yourself: take a bowl of gelatin, add some pineapple, and watch the gelatin dissolve. Chemistry in action!

PROJECT:
JUICY GELATIN DOTS

Modernist cooks sometimes call these little beads of gelatin "caviar," because they look like clumps of fish eggs, but they don't have to taste fishy. Try fruit juice, herbal tea, almond milk, root beer—then, think up ways to use different flavors in gel dot form. Creating them is easy—just dribble some gelatin in a tall glass of cold oil. As the blob of gelatin slowly sinks, it cools and solidifies. You'll also need a food-grade squeeze bottle and a strainer.

MATERIALS

Vegetable oil, such as corn or canola

4 teaspoons (20 mL) powdered unflavored gelatin (or two ¼ ounce packets)

3 tablespoons (45 mL) cold water

3 fluid ounces juice or other liquid

Ice

FIGURE 2-4: **Gel spheres flavored with fruit juice—in this case, pomegranate**

1. Fill a tall wide-mouth jar or drinking glass with oil, leaving some space at the top. Cover the jar and chill the oil in the refrigerator overnight or in the freezer for half an hour.

2. In a medium bowl, stir the gelatin into the cold water until it is smooth. Let the mixture stand and solidify.

3. Meanwhile, in a small microwave-safe container, heat the juice on high for 25 seconds until it is hot, but not boiling.

FIGURE 2-5: **Gelatin often comes in packets.**

Heat it again for 5–10 seconds, if necessary. Carefully pour the hot juice over the set gelatin mixture. Break up the gel with the spoon and stir until the gelatin is completely melted. Let it cool for several minutes until it is warm, but not scalding hot.

4. While you're waiting, prepare a bowl of ice water to keep the jar of oil cold when you take it out of the refrigerator. You can also practice using the squeeze bottle with a little water so that only a drop or two comes out at a time.

5. Time to make the dots! Pour the warm gelatin into the squeeze bottle. Put the jar of chilled oil in the bowl of ice water and remove the cover. Slowly let three or four drops of gelatin flow out of the bottle into the oil, one on top of the other. The drops should combine to form a ball and begin to sink to the bottom of the oil. Don't make the dots too big, or they will flatten when they hit

FIGURE 2-6: **Carefully fill the squeeze bottle.**

the bottom of the bottle.
It may take you a few tries to
get them just right.

6. Continue making little gel
spheres until you run out of the
gelatin mixture. Do them all at
once, because after a while the
gel will start to harden in the
bottle. If that happens, rinse
out the bottle using hot water,
mix up another batch of gelatin, and start again. Make sure
there is no gelatin stuck in the
nozzle.

FIGURE 2-7: **Controlling the amount of gelatin that comes out takes practice.**

7. As the oil jar fills up, use a
spoon to scoop out the dots.
Transfer them to a strainer to
drain. To serve the dots right
away, rinse off the last bit of oil
with cold water. To store the
dots for later, place them in a
container with a tight-fitting
lid. Cover them with a little
more oil, if needed, to keep
them fresh, and rinse them
right before using. Serve the
dots as a salad garnish or an ice
cream topping, or plop them
into a glass of soda and watch
them bob up and down with
the bubbles.

FIGURE 2-8: **A jar filling up with dots**

FIGURE 2-9: **Close-up of gelatin dots of vari-
ous sizes in a strainer**

Amaze your friends with a little magic trick! Make some grape juice dots and put half of them into a glass of water mixed with a teaspoon of baking soda. Slip the other half into some lemonade or lemon-lime soda. After a little while, the baking soda dots will turn deep blue, and the lemony dots will become reddish. Try switching some of the dots to the opposite glass to see how long it takes for them to reverse colors.

What's behind this magic? Chemistry! You probably already know that water molecules are made of two atoms of hydrogen (H) and one atom of oxygen (O), which is why water is referred to as H_2O. Sometimes, some of the hydrogen splits off from the water, leaving a piece of itself behind, to form a hydrogen ion (H+). The measure of the hydrogen ions in a liquid is called its *pH*. *Acids* have a low pH. They include lemon juice (which contains citric acid), vinegar (which has acetic acid), and cream of tartar (potassium bitartrate). Substances with a high pH, such as baking soda (sodium bicarbonate), are known as *bases*. Water itself and most foods are in the middle, or *neutral*. But many recipes use acidic ingredients to add a little zing.

Now, it just so happens that the chemical that gives grapes their color is a *pH indicator*. That means its chemical structure changes depending on whether it is mixed with an acid or a base. The change is slight, but with some pH indicators it is enough to create a whole range of different colors. Try adding more or less lemon juice to see if you can vary the shade of pink or blue of your gel dots!

FIGURE 2-10: Grape juice dots change color depending on how high the acid level is in the liquid.

PROJECT:
AGAR NOODLES

FIGURE 2-11: **Tomato-flavored gel noodles, garnished with mozzarella cheese balls and basil leaves**

MATERIALS

½ cup (125 mL) tomato juice or other flavored liquid

Scant ½ teaspoon (2.5 mL or 1 g) powdered agar

2 feet (60 cm) narrow silicone tubing—⅛ inches (3 mm) inside diameter, ¼ inches (6 mm) outside diameter—or plastic drinking straws

Large syringe or small squeeze bottle (Information on where to find the syringe can be found in the "Edible Pantry" materials list in the Introduction)

Jellied "spaghetti" made with agar is another favorite modernist invention. Get creative with your flavor combinations! Add mint leaves to garnish a plate of mint-tea-and-honey noodles. Serve tomato juice noodles with balls of fresh mozzarella cheese and basil leaves. Put a few maple syrup–flavored noodles on your pancakes, or make chocolate sauce, cherry, and whipped cream noodles for your ice cream sundae. Remember to take pictures and write down your recipe inventions, so you can make them again and share them with others.

FIGURE 2-12: **Tomato juice and agar are the only ingredients.**

1. In a small, microwavable cup, heat the juice in the microwave oven on high for 30 seconds. Stir in the agar powder. Put the cup back in the microwave for another 30 seconds, then take it out and stir again. Repeat four or five times until the mixture is clear. Watch to make sure the liquid doesn't boil over in the microwave. Let it cool for about five minutes. It should be warm, but not scalding hot.

2. Meanwhile, fill a bowl with ice water to cool the tubes when they are full.

3. When the agar mixture is cool, take the syringe and put the tip into the liquid. Slowly draw the plunger out, pulling juice into the syringe until it is full. (If you are using plastic drinking straws, see the directions in the "Drinking-Straw Noodles" sidebar.)

FIGURE 2-13: **Suck the liquid into the tube.**

4. Attach the end of a rubber tube over the tip of the syringe. Tie the tube into a loose coil. Be sure to point the other end of the tube away from you! Now push the plunger in and slowly fill the tube with juice.

5. Carefully remove the tube when it is full. Set the tube, still tied in a coil, into the bowl of ice water so the ends are pointing out of the water. Let it sit for about three minutes. If you have more tubes, keep filling them and placing them in the ice water.

FIGURE 2-14: **Loosely knot the tube to make it easier to handle.**

6. To remove the gel noodle from the tube, empty any leftover

FIGURE 2-15: **The gel firms up quickly.**

juice out of the syringe. Pull the plunger back so the syringe is full of air. Re-attach the tube to the syringe. Over a serving plate, slowly push the noodle out of the tube. Point the end of the tube where you want the noodle to go. You can make straight lines, wiggly designs, or spirals.

FIGURE 2-16: **Point the tube toward a plate—it slithers out fast.**

FIGURE 2-17: **Experiment with different colors and flavors.**

DRINKING-STRAW NOODLES

Food-grade, silicone-rubber tubing, which can stand up to high temperatures, is not easy to find in local stores. (See the "Edible Pantry" list at the front of the book for online sources—vinyl tubing from the hardware store is not recommended.) In a pinch, you can make shorter, thicker noodles using ordinary plastic drinking straws. As a bonus, instead of chilling them individually in a bowl of ice water, you can pop them all in the refrigerator at once. Here's how:

1. Place some drinking straws upright in a tall, straight jar or drinking glass, and set them in the refrigerator as you prepare the gel.

2. Use a syringe or squeeze bottle to pour a little agar mixture into each straw. Let it cover the bottom of the jar with a thin layer of gel. Return the jar to the refrigerator for a few minutes until the agar starts to set. The purpose is to plug the bottom of the straws enough to prevent all the gel from rushing out when you fill them.

3. Fill each straw with the gel, leaving a tiny bit of space at the top. Put the jar back in the refrigerator for about half an hour.

4. To remove the noodles, pinch the top end of the straw and start to push the noodle out the bottom onto your serving plate. Slide your fingers down the straw, squeezing as you go, until the noodle is completely out.

FIGURE 2-18: **Drinking straws make shorter, thicker noodles.**

FIGURE 2-19: **Use a smooth motion to squeeze the noodle out to avoid breaking it.**

Make any gel glow under a black light by using *tonic water*—a kind of soda that contains a fluorescent ingredient called *quinine*—instead of water. To prevent bubbles in your noodles or dots, let the soda go flat first by leaving it open until it reaches room temperature.

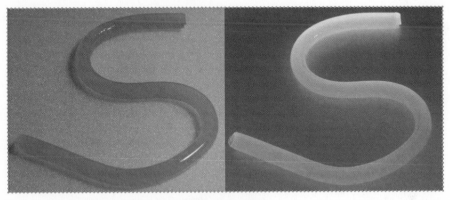

FIGURE 2-20: **Normal lighting and ultraviolet light**

RECIPE: AGAR-AGAR RAINDROP CAKE

MATERIALS

⅔ cup (150 mL) mineral water

Pinch sugar

⅛ teaspoon (0.5 mL) agar powder

Suggested toppings: honey, molasses, chocolate sauce, powdered ginger, brown sugar, crushed nuts, silver sprinkles

Optional: Maraschino cherry with stem, strawberries, or other colorful fruit

FIGURE 2-21: **A bamboo plate helps the presentation of raindrop cake.**

A *Raindrop Cake* is a clear gel dome so delicate that it turns to water in your mouth—and becomes a puddle on your plate if you don't eat it within 30 minutes! In Japan, it is called *mizu shingen mochi*, which means *water cake*, and is made with mineral water from mountain springs. Raindrop cakes are usually served on bamboo plates with a drizzle of black sugar syrup and a dusting of roasted soybean powder, but go ahead and use your favorite flavorings.

To create the round shape, you need a mold. Try a round-bottomed drinking or wine glass, a small rounded bowl, or a giant ice-ball mold. This recipe is enough for two palm-sized spheres.

1. In a small, microwavable cup, combine the water and sugar and heat the mixture in the microwave oven on high for 30 seconds.

2. Stir in the agar powder. Put the mixture back in the microwave for another 30 seconds, then take it out and stir again. Repeat four or five

times until the liquid is clear. Watch to make sure that it doesn't boil over in the microwave.

3. Pour the liquid into your mold. Gently pop any bubbles with a spoon. Chill it in the refrigerator for several hours or overnight. It should be solid to the touch, but still wobbly

4. To serve, gently tip the raindrop cake onto a plate so the rounded side is facing up. Serve with one of the toppings in the Materials list, or create your own!

5. *Variation:* To add a piece of fruit inside the raindrop cake, fill the mold halfway and put it in the refrigerator to set for about five minutes. Then place the fruit on top, and cover it with the rest of the liquid agar mixture. Let it chill until solid.

FIGURE 2-22: **If using a plastic wineglass, let the gel cool for a few minutes before pouring it in.**

FIGURE 2-23: **A piece of fruit in the raindrop cake makes it even more fun.**

FOAMY GOODNESS

Foams are created by mixing air or another gas into a liquid or solid. This can be done by simply stirring a liquid very fast for a long time with a fork or wire whisk, or beating it with an electric mixer. In molecular gastronomy, it's usually done by injecting nitrous oxide gas directly into the food. The tool used is a whipping siphon, which works a little like a can of whipped cream, but can be filled with any kind of liquid.

You can also produce foam using carbon dioxide gas. It's what gives carbonated drinks like soda their fizz, and creates a foamy head on the top. Some drinks are carbonated by adding yeast, a microscopic fungus that gives off carbon dioxide as it grows. Yeast also helps bread rise—in fact, bread is a kind of foam! But the most fun way to make carbonated foam is by creating a chemical reaction, a process called *effervescence*. Remember acids and bases? When you mix them together, you get a chemical reaction. If you combine baking soda (known chemically as sodium bicarbonate) with an acid like vinegar or lemon juice, the chemical reaction produces carbonation. Watch out, because that kind of reaction can quickly spill over! (It's how science fair volcanoes are made.)

Foams have different textures, depending on the size of the air pockets. You can see this yourself as you make the following meringue recipe. First you'll get a light and airy froth, like a bubble bath. In the soft-peak stage, the meringue will resemble shampoo lather. Finally, it will reach the stiff-peak stage and become as moldable and smooth as shaving cream. Also, notice how the color changes from almost clear to bright white. That's because all the bubbles bounce light around and keep you from seeing through to the other side.

FIGURE 2-24: **A tasty twist on the usual science fair volcano**

RECiPE: BAKED FOAM MERiNGUE COOKiES

MATERIALS

2 egg whites at room temperature
(see the "How to Use an Empty Water
Bottle as an Egg-Separating Device"
sidebar)

Optional: ¼ teaspoon (2 mL) cream of
tartar (you can substitute lemon juice
or white vinegar)

½ cup (125 mL) sugar

Optional: ½ teaspoon (4 mL) vanilla
extract, or other flavoring like almond
or mint

Optional: Food coloring

Parchment paper, or aluminum foil and
spray oil

You can make sweet, crispy cookies without any flour just by whipping egg whites into a foam called a *meringue*. But egg whites can be finicky. If the conditions aren't just right, they'll become gummy instead of foamy. They don't like it when the air is damp, and they really don't like it when there's

FIGURE 2-25: **Meringues are a perfect blend of lightness and crunch.**

any protein around. But there are some tricks that will make them perk up into nice stiff peaks. First, make sure the eggs are room temperature. (Warm them in a bowl of warm water for a couple minutes if you don't have time to leave them sitting out.) Adding sugar to the egg whites partway through the whipping process makes the "skin" of the bubbles thicker and less likely to pop. And adding an acid like cream of tartar makes it hard to overbeat the meringue. The acid prevents chemicals in the egg from sticking together and turning crumbly instead of puffy.

1. Preheat the oven to 225°F (105°C).

2. Beat the egg whites with the electric mixer on medium speed. Keep going until the eggs look frothy and bubbly. If you are using cream of tartar or another acid, add it now.

   Continue to beat the egg whites until they form soft, rounded peaks.

FIGURE 2-26: **The frothy stage**

3. Increase the speed to medium-high, and start to add the sugar. Pour in just a little at a time while the mixer is going. Continue to beat the egg white mixture until it gets very thick. Turn off the mixer and test whether the egg whites are ready. You will know they are done when they form stiff peaks that stand up on their own when you pull the mixer blades out.

FIGURE 2-27: **Cream of tartar is a powdered acid that doesn't change the taste of the mixture.**

FIGURE 2-28: **Soft peaks look like shampoo lather.**

FIGURE 2-29: **Beat in each scoop of sugar before adding the next.**

FIGURE 2-30: **Stiff peaks look shiny.**

4. If you like, fold in some vanilla extract or other flavoring. You can also add a drop or two of food coloring. If you divide the batch of meringue into separate bowls, you can make two or three colors and flavors at a time.

5. Line a cookie sheet with oven-proof parchment paper (or aluminum foil covered in spray oil). Drop teaspoon-sized blobs of the egg white mixture onto the pan, leaving space between them. As you shake the mixture off the spoon, make a swirly point at the top, like a chocolate kiss.

6. Bake for 45 minutes. Then turn the oven off, but leave the cookie sheet inside! Let the meringues sit in the oven until they are cool (for at least one hour). Several hours is even better. This lets

FIGURE 2-31: **Fold in the coloring gently with the flat side of a rubber spatula.**

FIGURE 2-32: **Give each cookie a twist as you plop it on the baking sheet.**

the center of the cookies keep baking until the whole meringue is crispy through and through.

FIGURE 2-33: Let the meringues cool for several hours for the best crunch.

THE INVENTION OF EGGLESS MERINGUE

In 2015, an Indiana software engineer named Goose Wohlt invented a way to make meringue without using eggs. His secret ingredient? The liquid from a can of chickpeas. Chickpeas contain a lot of protein, a nutritious substance found in meat, milk, eggs, and many kinds of beans. Protein is what gives meringue enough structure to hold together when it is whipped into a foam. Turns out that gummy liquid from the chickpea can (or cooking pot, if you make your own) absorbs enough of the bean's protein to work as an egg white substitute. Wohlt named his invention *aquafaba*, from the Latin words for water and bean. You can also use it to replace eggs in recipes such as chocolate mousse and mayonnaise.

HOW TO USE AN EMPTY WATER BOTTLE AS AN EGG-SEPARATING DEVICE

The time-honored method of separating the white from the yolk of an egg involves carefully cracking the egg in half and sliding the yolk back and forth between the two pieces of shell. This invention is much quicker and less messy. Do each egg separately, so if one yolk breaks, you don't spoil all the whites.

1. Break an egg into a bowl. Be careful not to break the yolk. (If the yolk does break, just set that egg aside and start again. You can refrigerate any leftover egg or egg white and use it in another dish later.)

2. Take an empty water bottle and squeeze it slightly to push out some of the air.

3. Hold the bottle over the yolk, and release your grip so the bottle expands out to its original shape. As the outside air tries to rush back into the bottle, the yolk will get sucked inside along with it.

4. Move the bottle over a separate bowl, and squeeze out the yolk. Add the white to your mixing bowl. Repeat until all the eggs are done.

FIGURE 2-34: **Squeeze the water bottle to create a vacuum.**

RECIPE: HOMEMADE WHIPPED GELATIN MARSHMALLOWS

MATERIALS

Spray oil

Powdered sugar (sift out any lumps by shaking or pushing it through a wire strainer)

1 packet or 1 tablespoon (15 mL) unflavored gelatin

¼ cup water

1 cup (250 mL) granulated table sugar

3 ounces (90 mL) water

Optional: ¼ teaspoon (2 mL) vanilla extract

Small square baking pan

Wax paper

Rubber spatula

FIGURE 2-35: Homemade marshmallows are soft and moist compared to the store-bought kind.

Once upon a time, there was a candy made out of mallow plants that grew in marshes. (Yes, really!) Today, marshmallows are made out of sugar and gelatin that has been whipped into a foam with an electric mixer. When you make your own, you can customize them any way you like—see the suggested variations following this recipe.

1. Spray a baking pan with oil. Sprinkle in some powdered sugar and shake it around to coat the entire inside of the pan.

2. Combine the water and gelatin in a measuring cup, and let it sit.

3. Combine the granulated sugar and the rest of the water in a saucepan. Stir the mixture over medium heat until the sugar is all dissolved. Pour in the gelatin mixture and bring the pot to a boil. Turn off the heat, take the pan off the stove, and let the pan sit to cool for a few minutes.

FIGURE 2-36: **The mixer stirs in air, making the gelatin and sugar puff up.**

4. Pour the mixture into a large mixing bowl. Beat it with an electric mixer until soft, about 10 to 15 minutes. You should see the mixture get thicker and increase in size. It's done when it looks shiny and white and it's thick enough for soft peaks to form.

5. Fold in the vanilla extract, if using. Pour the mixture into the pan and spread it around evenly. Cover the pan with wax paper and let it sit for several hours, or overnight. The longer it sits, the easier it will be to cut.

FIGURE 2-37: **Use the flat side of the spatula to gently pour the mixture into the pan.**

6. Lay a sheet of wax paper on the counter. Cover it with a thin layer of powdered sugar. Slide the tip of the rubber spatula under the edge of the marshmallow all the way around to loosen it from the pan. Carefully slide it onto the wax paper. Cut the marshmallow into squares.

A serrated table knife works well. Take each piece of marshmallow and flip it around so every side is dipped in the powdered sugar. This will keep it from sticking. Tap or pat each piece until you can barely see the powder. Your homemade marshmallows should keep in an airtight container for two or three weeks.

FIGURE 2-38: **If you let the marshmallow sit, it won't be sticky when you cut it.**

Variations:

- Add other flavorings or food coloring to the marshmallow mixture. Mint-flavored green marshmallows are one yummy combination.

- Instead of spreading the mixture in a pan, drop it by the spoonful onto on a parchment-lined cookie sheet, like the meringues earlier in this chapter. Make your own chick or bunny shapes!

- Instead of cutting it into squares with a knife, use cookie cutters in assorted shapes.

- If you love the popular marshmallow delivery system known as the S'more, make your marshmallows thinner and cut them into squares the same size as half a graham cracker. Use them in the following recipe for Marshmallow SunPies.

RECIPE: MARSHMALLOW SUNPIES

This chocolate and marshmallow sandwich on graham crackers isn't quite a s'more. Unlike the traditional campfire version, the marshmallow isn't toasted— homemade marshmallows melt when heated. And it's not really a MoonPie, because the chocolate is inside the graham cracker instead of covering the cookie. To

FIGURE 2-39: Cut your homemade marshmallows into flat squares for s'mores.

make it a true "SunPie," soften the chocolate in a Solar Oven (see Chapter 5) instead of the microwave.

1. On a microwave-safe plate, break a graham cracker in half. Break a chocolate bar in half, and put each piece on top of one of the graham cracker pieces.

2. Heat the chocolate and cracker in a microwave oven for about 20–25 seconds. The chocolate should be soft to the touch, but not runny.

3. Lay a square of marshmallow on one of the halves. Take the other half and flip it over and on top of the marshmallow, chocolate-side down. Squish the sandwich a little to get the chocolate to stick to the marshmallow. Enjoy!

FIGURE 2-40: **Don't overheat the chocolate.**

FIGURE 2-41: **Add the marshmallow, and press gently.**

RECIPE:
FIZZY WATERMELON LEMONADE

MATERIALS

Approximately 1 cup (250 mL) chilled watermelon chunks (seedless if possible)

Juice of 1 or 2 lemons

Optional: Sugar or honey to taste

½ teaspoon (2 mL) baking soda

Adding baking soda to lemonade makes a carbonated drink that's tangy and refreshing! You may need to adjust the amounts to get just the right amount of foam. To catch any spillover, put your glass of watermelon lemonade on a plate before adding the baking soda.

FIGURE 2-42: Fizzy watermelon lemonade is great on a hot day.

FIGURE 2-43: **You can use precut watermelon chunks or cut your own.**

1. Mash and mix up the watermelon chunks and the lemon juice in an electric blender until they are smooth. It should only take a few seconds. Taste the mixture before going on to see if it's sweet enough. If not, add a little sugar or honey.

2. Now get ready to add the bubbles! Stir in the baking soda and watch the carbonation happen. Wait until the foaming slows down before drinking your concoction.

FIGURE 2-44: **Don't forget to put the top on the blender before turning it on!**

3. *Variation:* Freeze the watermelon chunks before using and make your lemonade into a slushy.

FIGURE 2-45: **The baking soda goes to work.**

FIGURE 2-46: **Just the right amount of fizz**

THE CHEMISTRY OF CRYSTALS

Drying is another technique often used in molecular gastronomy. Modernist chefs use a dehydrator—a low-temperature heater that sucks all the moisture out of fruits, vegetables, and other foods in a matter of hours. They also dehydrate liquid mixtures to produce stretchy transparent films that are used to make Disappearing Ravioli, Flavored Handkerchiefs, and other bizarre-sounding dishes.

Under the right conditions, letting a liquid mixture dry out can also produce shiny crystals. A crystal is made up of microscopic molecules arranged in a repeating pattern. If you dissolve a substance like table sugar in water and let the water evaporate, or dry up naturally, the sugar will start to join up with other sugar molecules and form sugar crystals. The slower the process, the bigger the crystals can grow. The result? Rock candy!

Rock candy crystals are traditionally grown by leaving a piece of string or a stick in a solution of sugar water over several hours or several days. You can help the process along by heating the water before you add the sugar. This allows the water to absorb more sugar than it can at room temperature. Chemists call this a *supersaturated solution*. As the water cools back down to room temperature, it loses some of its ability to dissolve sugar, and the sugar molecules begin to "fall out of solution" and look for other sugar molecules to bond with. In this recipe, you will jumpstart the process by "seeding" a stick with dry sugar, which gives the loose sugar molecules something to grab onto as they fall out of solution.

FIGURE 2-47: **Rock candy takes patience, but it's worth it.**

PROJECT: ROCK CANDY STICKS

To make really great rock candy, you need to take your time. Growing the crystals to edible size can take from a few days to a couple of weeks, depending on how supersaturated your solution is. If you add too much sugar, your crystals will grow faster, but they'll be thin and spiky. You may also get crystals forming on the sides and bottom of the glass you're growing them

FIGURE 2-48: **The slower the rock candy forms, the larger the crystals are.**

in. If that happens, you can clean out the glass when you're done by filling it with hot water and letting it soak until the crystals dissolve again.

1. Pour the water into the jar and heat it in the microwave oven for one minute on high.

2. Remove the jar from the microwave. Slowly pour in the sugar and stir well.

3. Heat the jar on high for two minutes. Using oven mitts or pot holders, carefully remove the jar and stir the mixture.

FIGURE 2-49: **You can add food coloring and flavoring for special effects.**

Then, heat it on high again for another two minutes. Stir it again until all the sugar is dissolved. The liquid in the jar should be almost clear. You can add a few drops of food coloring to the mixture at this point. Allow the jar to sit and cool until it is warm or room temperature.

4. While you are waiting, prepare the sticks that your rock candy will be growing on. Take the coffee stirrers and dip them in the jar to coat the part where you want crystals to grow with the sugary syrup. Pour a little sugar on a plate. Roll the stirrers in the dry sugar until they are coated. Allow them to dry.

5. You can also make a holder to keep your coffee stirrers in place while the crystals grow. Stack several craft sticks together and wrap the small rubber band around one end to hold them together. When the sticks are dry, slip the uncoated end between the craft sticks. You may be able to fit more

FIGURE 2-50: **Don't let the sticks touch.**

than one, depending on how wide your jar is. When the sticks are all arranged, wrap the other rubber band around the other hand to hold them tight.

6. When the sugar mixture has cooled enough, take the holder with the coffee stirrers and set it on the rim of the jar so the sticks are in the liquid. The coffee stirrers should not be touching the bottom or the sides. Move them up or down in the holder as needed.

7. Cover the jar loosely with a coffee filter that will let the water inside evaporate without letting dirt or bugs in. If bugs or other critters are a real concern, wrap the paper tightly around the opening of the jar with the canning jar lid, a rubber band, or tape. (You may be able to get the sticks to stay in place just by poking them through the paper if it is tight enough, so you don't need to use a holder.)

8. You may start to see crystals grow after one day, but if you leave it for several days, the layers of crystals will grow. The slower your crystals develop, the larger they will be. When your candy sticks are a good size, remove them from the liquid and let them drip dry in an empty jar or cup. Eat them now or wrap them in plastic and save them for a special treat!

FIGURE 2-51: **Finished sticks can be wrapped as gifts.**

FLASH-FROZEN DELIGHTS

When modernist cooks make frozen desserts, they up the coolness factor by pouring on super-cold liquid nitrogen. At normal temperatures, nitrogen is a gas that makes up most of the air we breathe. To transform nitrogen gas into a liquid, you have to lower its temperature to -321°F (-196°C). That's quite a bit colder than the typical home freezer, which is usually set at 0°F (-18°C). Liquid nitrogen can freeze ingredients so fast that ice crystals don't have time to form, making desserts extra smooth and creamy. But it is hard to find (outside a laboratory), and dangerous to handle. Luckily, you can get nearly the same results with dry ice, which is the name for solid carbon dioxide. Compared to your freezer, it's still pretty cold at -109°F (-78.5°C), but it's much easier to find. You can buy dry ice at some supermarkets and beverage centers, as well as at welding shops. Ten pounds (4.5 kg) should last about 24 hours before it sublimates back into carbon dioxide gas.

FIGURE 2-52: **Dry ice makes water "boil" at room temperature.**

SAFETY RULES FOR WORKING WITH DRY ICE

Dry ice is still cold enough to cause frostbite injuries, and in closed spaces, the carbon dioxide gas can build up, so be sure to follow these safety rules:

- Store dry ice in a container with a loose lid, like a Styrofoam cooler. Never put dry ice in a tightly closed container, especially one made of glass. The pressure of the carbon dioxide gas released as the dry ice warms can cause the container to shatter.

- Always wear insulated gloves or oven mitts when handling dry ice. Avoid touching it with your bare skin. It can cause frostbite injuries in just a few seconds. Also protect yourself when handling a bag or non-insulated container holding dry ice, or when using non-insulated tools to handle the dry ice.

- It's a good idea to wear eye protection, like safety goggles, when handling dry ice.

- Never put dry ice directly in your food! Swallowing a piece of dry ice could cause severe injuries.

- Try to use plastic tongs, scoops, and other tools when moving dry ice around. Metal tools transfer heat to the dry ice and make it sublimate faster. It's not dangerous, but the pressure of the carbon dioxide gas against the metal tool can cause it to produce a screaming sound!

- Only use dry ice in a space with good air flow so the carbon dioxide gas it gives off can escape. Carbon dioxide isn't poisonous—you produce it every time you exhale—but in a closed space it can build up and make it hard to get enough oxygen to breathe. Avoid breathing in the "fog" for the same reason.

RECiPE:
DRY iCE SORBET

MATERIALS

1 cup (250 mL) of pureed fresh fruit, one or more kinds

¼ cup (60 mL) sugar

Sorbet is just fruit that's frozen. It stays slushy instead of freezing solid thanks to sugar molecules. They get in the way of the water molecules, so the water can't form ice crystals. The water stays cold but liquid-y. The amount of sugar you need to add to achieve the right degree of slushiness varies, depending on how sweet the fruit is. Since every batch of fruit is different, you'll need to experiment by adding more or less sugar as you go.

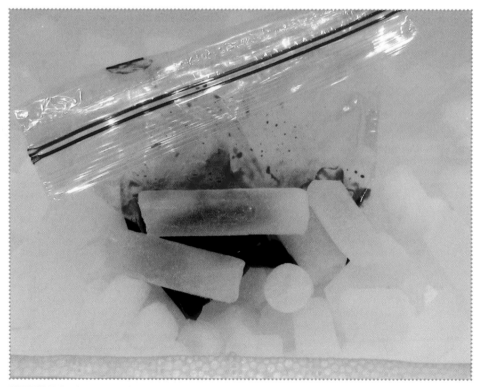

FIGURE 2-53: **Use sandwich bags to make quick, single servings of frozen fruit puree.**

The easiest fruits to make into sorbet include berries, bananas, and mango. Fresh is best, but you can use frozen fruit, as well. Just let it soften first. To start, you'll need to mash up or puree your fruit using a fork, potato masher, blender, or food processor. If you want it smoother, push it through a strainer before using. If it is too thick, add a little water.

FIGURE 2-54: **Try different combinations of your favorite fruit.**

FIGURE 2-55: **Many fruits are soft enough to puree without a blender.**

FIGURE 2-56: **Mango and bananas are easy to mash with a fork.**

FIGURE 2-57: **Sugar helps keep the fruit from freezing solid.**

1. Puree the fruit with a potato masher, fork, electric blender, or food processor. Add the sugar and place the mixture in a zip-top bag. Flatten the fruit mixture inside the bag.

2. Arrange the dry ice inside the cooler or in a bowl so that you can lay one or more bags on top. Make sure the bag is completely sealed, and lay it on the dry ice for one minute.

FIGURE 2-58: **Protect your hands from frostbite by wearing ski gloves or oven mitts.**

3. Remove the bag carefully. Wearing gloves, squeeze the bag to see if it is frozen enough. If there are spots that are too soft, squish the icy patches around to mix them in. Then, lay the bag back on the dry ice for another minute. Repeat until you're happy with the texture of the frozen fruit mixture.

4. You can eat your sorbet right from the bag, or spoon or squeeze it into a cup or bowl.

5. *Variation:* Turn your sorbet into sherbet by replacing half the fruit with whole milk. For an orange creamsicle flavor, use orange juice instead of fruit puree.

FIGURE 2-59: **Eat your sorbet right out of the bag, or serve it in a cup.**

ROJECT:
KE BELOW-FREEZING
iCE SORBET

FIGURE 2-60: **Mixing salt in with the ice lowers the temperature for faster freezing.**

MATERIALS

1 cup (250 mL) sorbet mixture (see earlier directions for dry ice sorbet)

1 sandwich-size, zip-top plastic bag

1 large zip-top plastic bag

2 cups (500 mL) ice

1 cup (250 mL) salt

1 cup (250 mL) water

Don't have dry ice? You can still make sorbet using ordinary ice water. The secret is to add salt. Just like the sugar in the sorbet mixture, the salt keeps ice crystals from forming, and allows the water to reach below-freezing temperatures.

> **Tip** Although the ice mixture won't reach dangerously-cold temperatures, you'll be more comfortable wearing gloves or oven mitts when handling it.

1. Pour the sorbet mixture into the sandwich-size plastic bag. Seal it tight. (For extra security, you can seal that bag inside another small plastic bag.)

2. Place the ice, salt, and water inside the large bag.

3. Put the small bag into the large bag along with the ice mixture.

4. Put on your gloves. Shake and squeeze the bag for a couple of m
   Then stop and feel the smaller bag (you can leave it inside the la
   bag) to see if it's ready. If the sorbet isn't frozen enough for you, shake it
   for another couple minutes. It should take about five minutes to get the
   sorbet to the texture you want.

Tip You can make ice cream using either of the previously described methods.
Just mix heavy cream, a little sugar, and some flavoring like vanilla extract or
chocolate syrup.

MORE ABOUT CHEMICAL CUISINE

Amazing Food Made Easy (amazingfoodmadeeasy.com)

Harvard University Science and Cooking videos (online-learning.harvard.edu/course/science-and-cooking)

Texture: A Hydrocolloid Recipe Collection edited by Martin Lersch (blog.khymos.org/wp-content/2009/02/hydrocolloid-recipe-collection-v3.0.pdf)

Culinary Reactions by Simon Quellen Field (Chicago Review Press, 2012) (kitchenscience.sci-toys.com)

Hack It from Scratch

3

Make do-it-yourself versions of your favorite ready-to-eat foods at home—with ingredients you already have in your pantry!

FIGURE 3-1: Homemade fries, ketchup, and pickles make convenience food like hot dogs even tastier.

For thousands of years, most households prepared all their meals from scratch every day. If not, they bought their meat, bread, and cheese freshly made from neighborhood shops. But even shopkeepers only prepared enough for local customers, because it was hard to store food or ship it to far-away places without it going bad.

Home cooks had always preserved extra food for their own use using techniques like drying, smoking, fermenting, and pickling. Then in the early 1800s, Emperor Napoleon Bonaparte of France offered a reward to anyone who could figure out how his troops could carry perishable food with them as he sent them off to conquer the world. A candy maker named Nicolas Appert had worked on the problem for many years. He won the prize with his idea to cook the food inside glass bottles that were sealed with corks covered with wax to make them air-tight. His invention inspired Englishman Peter Durand, who developed the tin can in 1810. The tin can was less breakable, but didn't really become popular until the can opener was invented 50 years later. In 1864, scientist Louis Pasteur discovered why canning worked—cooking food in sealed containers killed off microscopic bacteria that caused it to spoil and kept new bacteria out. His discovery made it possible to find the best temperatures for safe canning.

When inventor Clarence Birdseye developed his Quick Freeze Machine to package fresh vegetables in 1925, American families began to stock up their new refrigerator-freezers. But frozen foods really became popular when the first complete precooked frozen meal came along in 1954. Gerry Thomas, a salesman for C.A. Swanson & Sons, came up with the idea after his company was stuck with too many leftover Thanksgiving turkeys. His invention was called the TV dinner, and it came on a single-serving alumi-num tray so you could eat it in the family room as you watched your favorite TV show. Swanson sold 10 million frozen TV dinners in its first year.

Pre-prepared foods are convenient, but that's not the only reason they're so popular. Food manufacturers tailor their products to appeal to the human craving for foods that are sweet, salty, and full of fat. Back when early humans had to grow or hunt their own meals, that preference helped them survive. But nowadays, eating too much sweet, salty, or fatty food can be bad for your health. Processed food also contains a lot of artificial ingre-dients, such as corn syrup, that make it cheaper to produce, but harder to digest, than food made from fresh ingredients.

That's why there's new interest in cooking from scratch. This doesn't mean you have to give up your favorite foods. In fact, making your own version of many popular convenience foods is almost as easy as opening up a package from the store. (It helps to keep ingredients on hand for when you get the urge to whip something up. See the "Edible Pantry" supply list at the beginning of the book.) Making something yourself also lets you control what goes in your food. When you're the chef, you can make it taste just the way you want. Experiment with the following recipes and see what new favorites you can invent for yourself and your friends and family!

 WASTE NOT, WANT NOT

Pre-prepared foods have another downside: they usually involve a lot more packaging than fresh ingredients. Think about all the plastic that goes into a box of individually wrapped snack cakes, as opposed to cupcakes you bake yourself. Of course, when you're using raw fruits and vegetables, you may be left with a pile of unused bits and pieces, but that's the kind of waste that can be good for the environment! So, don't throw the peels and ends in the trash. In the next chapter, you'll learn how to recycle leftover fruit and vegetable scraps—and even how to use them to grow whole new plants!

THE HiSTORY OF COLD BREAKFAST

Breakfast cereal was one of the first convenience foods ever invented, as well as one of the first "health foods." In the late 1800s, a doctor named John Harvey Kellogg ran a rest home in Battle Creek, Michigan, that served vegetarian food as a cure for stomach ailments. A patient, C. W. Post, enjoyed the breakfast grains they served there so much he began to sell his own version under the name Post Toasties.

FIGURE 3-2: **Granola was one of the first commercial breakfast cereals.**

Not to be outdone, Dr. Kellogg's brother started his own company to sell Kellogg's Corn Flakes. Soon, dozens of cereal companies sprang up in Battle Creek, each offering their own ready-to-eat breakfast fare. Meanwhile, in Niagara Falls, New York, an engineer named Henry Perky came up with the idea for shredded wheat. Thousands of tourists visiting the falls for their honeymoons or vacations made sure to visit the Nabisco Shredded Wheat factory while they were there.

Granola was invented in 1863 by Dr. James Caleb Jackson to go along with the mineral-spring treatments at his own health resort in upstate New York. His version was a little dry, but his idea was copied and improved by the Kelloggs. By 1889, the Battle Creek cereal barons were selling two tons of granola a week. Today, there are many different ways to make granola, and you can create your own.

RECIPE: CRUNCHY GRANOLA

MATERIALS

4 cups (1 L) rolled oats (sometimes labeled as "old fashioned")

½ cup (125 mL) sunflower seeds

⅓ cup (80 mL) sesame seeds

½ cup (125 mL) sliced or crushed almonds

Pinch salt

¼ cup (60 mL) vegetable oil or butter

¼ cup (60 mL) honey

½ teaspoon (2.5 mL) vanilla extract

1 cup (250 mL) raisins

The main ingredient is usually rolled oat flakes, but you can use a combination of different grain flakes. You can also use pumpkin seeds instead of

FIGURE 3-3: Homemade granola, hot out of the oven

sunflower seeds, and any kind of dried fruit and nuts instead of almonds and raisins, such as walnuts and dried cranberries.

1. Preheat oven to 300°F (150°C).

2. Combine all the dry ingredients in a large bowl. Stir until well mixed.

3. In a microwave-safe measuring cup or bowl, combine the oil, honey, and vanilla. Heat the mixture on high for 30 seconds, then stir with a fork. Repeat if needed until the mixture is runny.

FIGURE 3-4: **Oats, sesame seeds, slivered almonds, and sunflower seeds**

4. Pour the oil and honey mixture into your dry ingredients. Toss using a large spoon until everything is well coated.

5. Spread the mixture on a baking pan, about ½ inch (1 cm) thick. Put the pan in the oven for 15 minutes. Then, take the pan out and stir the mixture around so it bakes evenly. Put it back in the oven for another 15 minutes. Then, take it out, add the raisins, stir it again, and let it bake for another 15 minutes.

FIGURE 3-5: **Honey and vegetable oil**

6. Your granola is done when it is brown and smells a bit like fresh-baked cookies. Take the pan out and let it cool. Store your granola on the shelf in an airtight container (or refrigerate if you used butter). Homemade granola goes great with homemade yogurt—see the recipe in Chapter 5!

THE iNVENTiON THAT CHANGED BAKiNG

Way back in the 1840s, baking involved some complicated chemistry. Whenever a housewife wanted to make biscuits or cake, she would go down to the drugstore and buy packets of two powdered chemicals you met in Chapter 2: sodium bicarbonate (baking soda) and potassium bitartrate (cream of tartar). When she got home, she measured the chemicals out as carefully as a scientist in a laboratory. When combined in the proper amounts, the acid and the base bubbled up to make baked goods light and fluffy. The process was quicker than waiting for yeast bread to rise, but it was still a lot of work.

Then, in 1869, a scientist named Eben Norton Horsford found a way to save bakers a step. His company, Rumford Chemical Works in Rhode Island, combined the two chemicals into one product that could sit on a shelf for long periods. Horsford's invention, baking powder, made convenience foods like pancake mix possible.

 SAFETY REMiNDER

Always get adult help when cooking or cutting in the kitchen. If you need a stool to reach the stove, make sure it won't tip. Keep oven mitts or pot holders around to grab hot pans, but be sure not to leave them near a hot burner. Young beginning cooks can chop up many fruits and vegetables with a rounded butter knife. Put the item you're cutting on a cutting board and keep the sharp edge of the knife pointed down—making sure there are no fingers in the way!

RECIPE: PANCAKE MIX

MATERIALS

4 cups flour (1 L) (white, whole wheat, or other varieties in any combination)

1 tablespoon plus 1 teaspoon (20 mL) baking powder

1 teaspoon (5 mL) baking soda

1½ teaspoons (7 mL) salt

1½ teaspoons (7 mL) salt

Optional

> 8 teaspoons (40 mL) of cream of tartar
>
> 6 teaspoons (20 mL) of baking soda

FIGURE 3-6: **Make your own pancake mix and save time in the morning.**

Pancake expert Amy Halloran, who created this recipe, still uses Rumford Baking Powder in her homemade pancake mix.

1. Mix ingredients well with a whisk or large spoon.

2. Store in a plastic bag or tightly sealed container. To use, see the following recipe.

3. *Extension:* Instead of using pre-made baking powder, mix up your own leavening chemicals, like housewives did in the old days. The formula: two parts cream of tarter to one part baking soda. To try it in this pancake mix recipe, start by combining just the flour and salt. Then, in a separate bowl, mix the cream of tartar and 6 teaspoons (20 mL) of baking soda, and stir well. Measure out 4 teaspoons of the mixture and add it to the flour and salt. You can save the leftover cream of tartar mixture for another batch of pancake mix--or dump a little in a glass of water and see what happens!

RECIPE:
PANCAKE MIX PANCAKES

FIGURE 3-7: **Pancakes in the skillet—yum!**

MATERIALS

1 egg

¾ cup (180 mL) milk

**1 tablespoon (15 mL) yogurt
(see the recipe in Chapter 5)**

**1 cup (250 mL) pancake mix
(from previous recipe)**

Butter

1. In a large bowl, beat the eggs lightly with a whisk or fork until completely yellow. Add the milk and yogurt and mix well.

2. Add the pancake mix. Use the whisk or a large spoon to combine the ingredients, but be careful not to over-stir! You should stop as soon as there are no more large clumps of dry ingredients. The batter should still be lumpy.

3. Let the batter sit for 10 minutes. This helps the flour absorb the rest of the moisture and start to rise. If the batter needs a little thinning, add some more milk.

4. About five minutes before you are ready to start cooking, heat a griddle or frying pan (Amy Halloran recommends aluminum for even heating) on low. Then, turn it up to medium. It is ready when a drop of water on the surface dances around instead of vaporizing. Grease the pan with some butter until it melts. As soon as the butter starts to turn light brown, use a ¼ cup measuring cup or other small scoop to pour blobs of batter onto the hot skillet. You can make the pancakes whatever size

you prefer, but smaller pancakes cook faster and are easier to flip with a spatula.

5. As the pancakes cook, the batter will start to bubble. This should only take about a minute! When most of the bubbles have burst, use a spatula to flip the pancake over. Let it cook on the other side for about 30 seconds. Then, transfer the pancakes to a serving plate. Eat immediately, or, if needed, keep them warm on a baking sheet in an oven set on low until you are ready to serve them.

6. Homemade pancakes are good served with maple syrup, honey, homemade applesauce (see the recipe later on in this chapter), or more yogurt. You can also drop berries, slices of fresh fruit, or chocolate chips onto each pancake right after you pour the batter onto the skillet.

FIGURE 3-8: **Flip the pancakes when the bubbles are just starting to burst.**

Tip   To use this pancake recipe for drawing pictures on the griddle (see Chapter 1), don't let the batter sit and rise. Your drawing will look better if the pancake is flatter.

CONDIMENTS AND SPREADS

Condiments and spreads are another type of food that became big business when companies figured out how to package them for maximum shelf life. Ketchup was one of the first bottled foods in America, but its history goes way back before the invention of bottling plants, and it hasn't always been made with tomatoes. The name comes from the Chinese *ke-tsiap*, a type of fish sauce. In Europe in the 1600s, walnut, oyster, and mushroom were popular ketchup flavors. Tomato ketchup appeared by 1812, but early bottled ketchup often spoiled or contained preservatives that made people sick. On top of which, the glass bottles would occasionally explode.

The ketchup situation was one of the reasons the U. S. government passed the Pure Food and Drug Act in 1906 and later created the Food and Drug Administration. Luckily for ketchup lovers, a man named Henry John Heinz, who began manufacturing ketchup in Pittsburgh, in 1876, started using sterilized bottles, better quality tomatoes, and vinegar, instead

FIGURE 3-9: **Do-it-yourself ketchup, ready to go in the jar**

of toxic ingredients, to kill harmful bacteria. His safety measures made ketchup an American classic.

Unlike some other kinds of food processing, pickling has been around for thousands of years. As with Heinz's ketchup, pickles use the acid in vinegar to kill off harmful bacteria and keep food from spoiling. Other types of pickles use saltwater to encourage helpful bacteria to grow and keep bad bacteria out. But that's not the only reason people pickle their food. The pickling process also adds sweet, sour, or sharp flavors, and changes the color and texture in delicious ways. In the U. S., most pickles are made with cucumbers, but in other parts of the world, you can find pickled cabbage, tomatoes, and even pickled duck eggs.

Fermentation is another way to prevent harmful bacteria from growing—in this case, by letting friendly, non-toxic bacteria start the digestion process for you. As the good bacteria grow, they produce carbon dioxide gas, which makes the food acidic (just like vinegar) to kill off harmful microbes.

 WHY DO i HAVE TO REFRIGERATE MY HOMEMADE FOOD?

The condiments you buy at the store can sit on a shelf for years without going bad, because they've been cooked in sterilized jars using high heat. When people can their own food at home, they use a similar process. But the recipes in this book don't use high heat and sterilized containers to keep food safe. So, even if you're using a canning jar to store them, you need to refrigerate your condiments as soon as they're done.

RECIPE: TANGY FERMENTED KETCHUP FROM SCRATCH

🥄 MATERIALS

3 6-ounces cans of tomato paste (or 2¼ cups (750 mL) homemade tomato paste)

⅓ cup honey

2 tablespoons apple cider vinegar

⅓ cup whey

Salt and pepper to taste

Optional:

    Cinnamon

Allspice

Ground cloves

Garlic

Celery salt

Worcestershire sauce

Cumin

Hot pepper (paprika, chili, chopped pepper flakes, or cayenne)

Making homemade ketchup that tastes like the commercial kind is near impossible, so you might as well make something unique—like fizzy, fermented ketchup. The fermentation comes from whey, the yellowish liquid you get when yogurt separates. Make sure the yogurt you use says "active cultures" on the label—that way you know it comes with live bacteria. Or make fresh yogurt using the recipe in Chapter 5. This recipe also saves time by using canned tomato paste. You can substitute fresh tomatoes that have been cooked and strained to make a super-thick sauce.

FIGURE 3-10: **Make your ketchup as thick as you like.**

1. Combine all ingredients in a medium bowl and mix well. If you want thinner ketchup, add a little more whey or water.

2. Spoon the mixture into the canning jar. Leave space at the neck of the jar.

3. Using a clean cloth or paper towel, wipe off any ketchup around the inside of the rim.

4. Cover the jar with a piece of cheesecloth, and secure with a rubber band or the rim of the canning lid. Let the jar sit at room temperature for 24 hours, or as long as five days. The longer it ferments, the fizzier it will get.

5. When you're happy with the ketchup, put the jar in the refrigerator. It may continue to ferment for a day or two in the fridge, so be sure to "burp" the jar by opening it for a second to let out any gas that has built up. Your fermented ketchup should keep in the fridge for several weeks. Serve with homemade French Fries (next recipe).

FIGURE 3-11: That yellow liquid is whey from yogurt—full of friendly microbes.

FIGURE 3-12: Bonus tip: The easiest way to get tomato paste out of those little cans is to cut off both the top and the bottom and push the paste out. Watch out for sharp edges!

RECIPE:
OVEN-BAKED FRENCH FRIES

MATERIALS

4 potatoes (or 1 per person)

Olive oil

Salt, pepper, and other seasoning to taste (try paprika or nutmeg)

FIGURE 3-13: **Fries cook up nicely in the oven.**

French fries are America's favorite vegetable. President Thomas Jefferson brought them back with him in 1789, after serving as U. S. minister to France. But they may have actually gotten their start in neighboring Belgium, which loves fries so much it gave them their own museum. In the 1970s, new machinery made it possible to peel, cut, cook, and deep freeze potatoes in factories, and French fries became one of the most popular fast foods. The homemade fries in this recipe are baked in the oven, so they use less oil than deep-frying, but they still taste great. Most commercial French fries use Idaho potatoes, but you can try different varieties. For yummy sweet potato fries, add a little nutmeg to the seasoning. You don't even have to stick to potatoes. Try this oven-roasting recipe with green beans, leeks, parsnips, beets, or other vegetables (although they may not come out as crisp). Add onions or cloves of garlic for extra flavor.

BEWARE OF OIL SPLATTERS!

To avoid burns from hot oil, be sure to handle pans with oven mitts or hot pads. An apron can help protect clothing from any oil that splashes.

1. Preheat oven to 450°F (230°C).

2. Peel potatoes, or leave the skin on but wash them well.

FIGURE 3-14: **Wash the potato skins well if you're leaving them on your fries.**

3. Cut potatoes into wedges or long strips. If you're using small potatoes, you can also slice them into rounds, like thick potato chips.

4. Lay the potatoes in a single layer on one or more cookie sheets. If the potatoes are moist, you can blot them with a paper or cloth towel. This will help make them crispier.

5. Drizzle olive oil over the potatoes until they are well covered. Flip them with a spatula to coat them with oil on both sides.

6. Let them cook for 20 minutes, more or less, depending on how thick the fries are. Halfway through, remove the cookie sheets, flip the potatoes over with the spatula, and return them to the oven. The fries are done when they start to brown around the edges. Take them out and season to taste. Serve with some homemade fermented ketchup, which you learned to make in the previous recipe!

RECiPE:
SWEET REFRiGERATOR PiCKLES

🥄 MATERIALS

3–5 pickling cucumbers

½ sweet onion

¾ cup (180 mL) granulated sugar

½ tablespoon (7 mL) kosher or pickling salt

1 tablespoon (15 mL) mustard seed

1 tablespoon (15 mL) dill (dried or chopped fresh)

1–2 cups (250–500 mL) white vinegar

These sweet cucumber pickles are super easy. They don't need to be cooked, and they are ready overnight! Look for Kirby pickling cucumbers at your local grocery store or farmer's market, or grow your own. They

FIGURE 3-15: **Sweet and crispy!**

are shorter and bumpier than regular salad cucumbers, and their skins are much thinner. This recipe also works well with Japanese cucumbers, which are longer and skinnier than regular cucumbers and usually curved.

1. Wash the cucumbers. You do not need to peel them. If they are fresh from the garden they may have tiny black "seeds" on the skin, which you can rub off when you wash them.

2. Cut off the ends and slice the cucumbers into rounds about ¼ inch (6 mm) thick.

3. Dice the onion (cut it into small square pieces). In a large bowl, toss them with the cucumber slices. Transfer the ingredients into a jar. Pack in as much of the cucumber and onion as you can, right up to the rim. A large-mouthed canning funnel makes it easier to avoid spilling any.

4. In a small bowl combine the sugar, salt, mustard seed, and dill. Add 1 cup of vinegar. Stir until the sugar is dissolved.

FIGURE 3-16: **Pickling salt is made up of tiny grains and has no additives that can make the water cloudy.**

FIGURE 3-17: **Dicing the onions means cutting them up into little squares.**

5. Pour the vinegar mixture into the jar. Add more vinegar, if needed, until the cucumbers and onions are completely covered. Close the top of your jar tightly and swirl it around a little. Then, leave it overnight in the refrigerator. By the next day, your pickles should be darker and somewhat softer, but still crunchy. They go great with sandwiches, hot dogs, and hamburgers.

RECIPE: HOMEMADE HUMMUS

MATERIALS

1½ cups (350 mL) canned or home-cooked chickpeas

3 tablespoons (45 mL) lemon juice

1 tablespoon (15 mL) olive oil

3 tablespoons (45 mL) tahini (sesame paste)—you can substitute peanut butter or yogurt

½ teaspoon (3 mL) cumin

Salt, pepper, other seasonings to taste

Optional:

   1 clove garlic, chopped

   3–4 scallions (green onions)

   3–4 stalks of parsley

   1 carrot, chopped

Hummus is a popular spread or dip made with chickpeas. Its roots go far back to Greece, the Arab world, and Israel, but ready-made hummus has only been readily available in supermarkets since the 1990s. Along with the traditional tahini (sesame paste), garlic, and lemon, you can now get hummus in a wide variety of flavors, from guacamole to chocolate.

FIGURE 3-18: **Hummus is great in salads and sandwiches, or as a dip.**

> **CHICKPEAS FROM SCRATCH**

You can use dried chickpeas instead of canned, but you have to soak and cook them first. Rinse the dried chickpeas and remove any tiny stones. Place them in a pot, and cover with two or three times as much water. Soak overnight (or boil for two minutes and let them sit for two hours). Then drain the water, refill the pot, and bring the chickpeas to a boil. Lower the heat and let the chickpeas slowly simmer for one hour. Next, drain them and let them cool, then use them just like canned chickpeas.

This recipe is for traditional hummus, but feel free to experiment. For something a bit sweeter, try making it with peanut butter! Hummus can be served as a dip with wedges of toasted pita bread or with vegetable sticks, or used as a spread on bagels. You can also stuff it in a pita pocket with radish or alfalfa sprouts. (See Chapter 4 for sprout-growing directions.)

> **SAVE YOUR AQUAFABA!**

Don't throw away that chickpea liquid! You can use it as an egg substitute for meringues and other recipes. See Chapter 2 for information on the liquid, which has been dubbed *aquafaba*.

1. If you're using canned chickpeas, drain the liquid out and rinse the chickpeas to remove any liquid that clings to them.

2. Mash the chickpeas in a bowl with a large spoon or potato masher, or use an electric food processor. Stir or blend in lemon juice and olive oil until you get a somewhat smooth texture.

FIGURE 3-19: **Use a strainer to drain the liquid from your dried or canned chickpeas.**

FIGURE 3-20: **Once you mash up the chickpeas with (or without) sauce, you can add your choice of veggies.**

3. For traditional-style hummus, stir or blend in some tahini. It has a strong taste, so if you prefer, you can also substitute peanut butter or yogurt—or just leave it out altogether. If your hummus is too thick and chunky, stir in a few spoonfuls of water to thin it out.

4. Add cumin and other seasonings, as desired. If you're using garlic or carrots, chop them up into tiny chunks and stir them in. For scallions and parsley, you can snip them over the bowl with clean kitchen shears.

NUKE iT! THE MiCROWAVE REVOLUTiON

FIGURE 3-21: **Microwave ovens have transformed the way we eat.**

Of all the food inventions in the twentieth century, the one that probably did the most to change the way Americans eat is probably the microwave oven. Its discovery goes back to 1946, when an engineer named Percy Spencer noticed that the radar machines he was working on had melted a candy bar in his shirt pocket. The first industrial microwave ovens were big and expensive, and used mainly in restaurants. By the 1970s, they had become small enough and affordable enough for home use. While they're not the best way to cook a lot of foods, today they're used by kids and adults as a quick, safe way to reheat leftovers or make a quick single-serving snack.

RECIPE: APPLESAUCE CAKE IN A MUG

MATERIALS

¼ cup (60 mL) all-purpose flour

3 tablespoons (45 mL) brown sugar

Pinch of salt

½ teaspoon (2 mL) apple pie spice, or the following mixture:

  ¼ teaspoon (1 mL) cinnamon

  ⅛ teaspoon (0.5 mL) ginger or cardamom

  Pinch of nutmeg

Pinch of allspice

Pinch of cloves

2 tablespoons (30 mL) unsweetened applesauce (see recipe that follows) or one apple, diced up into small pieces

2 tablespoons milk

1 teaspoon vegetable oil (or 1 teaspoon warm softened butter)

1 teaspoon water

Microwave-safe mug, 8 ounces or larger

When you try to bake in the microwave oven, you run the risk of ending up with something soggy. But microwaved cake-in-a-mug comes out moist, without being gooey.

FIGURE 3-22: **Single-serving cake, warm from the microwave, in just about two minutes**

1. In a mug, mix up the dry ingredients with a fork until well blended.

2. Make a *well*—a small shallow hole in the middle of the dry ingredients. Pour the wet ingredients into the well. Stir with a fork until the batter is completely mixed.

FIGURE 3-23: **Experiment with different spices.**

3. Put the mug in the microwave and heat on high for two minutes. Use a fork to gently slide the cake out of the mug onto a plate. Careful; it will be hot and steamy! If the sides or bottom of the cake still seem sticky, you can slide it back into the mug and heat it for another 20 seconds.

4. Use your fork or a knife to slice the cake open. Let the steam escape before eating. Microwave cake is especially good topped with a little whipped cream.

FIGURE 3-24: **When it's finished, you can slide your cake out of the mug.**

RECiPE:
HOMEMADE APPLESAUCE

FIGURE 3-25: **It doesn't take much to make delicious applesauce.**

Making applesauce on the stovetop is so fast and easy, it seems a shame to use the store-bought kind. The hardest part is peeling all the apples, so this recipe saves you a step by leaving the apple skin on while it's cooking! The peel gives the applesauce a slight pink tinge, but it tastes just as good.

1. Wash the apples and let them dry. Take the first apple and cut it into four parts by slicing off one side at a time, leaving a squared-off core at the center. Dispose of the core, and put the slices into a large saucepan. Squirt some lemon juice on the slices—the acid in the juice will keep the slices from turning brown. Repeat with the other apples.

2. Add the water to the apple slices in the saucepan. Cook the apples over medium heat, stirring every few minutes with a large wooden spoon. As the apples soften, break them up with the wooden spoon. In

FIGURE 3-26: **Cut off the sides of the apple and discard the core.**

about 15 minutes, they should be completely mushy and start to separate from the peels.

3. Let the apples cool for a few minutes. Then, break up any slices that are still holding together with the wooden spoon. The skin at this point should have separated from the apple slices. Use a fork to dig around in the sauce and pick them all out.

FIGURE 3-27: **Add some water to the apples to help them break down faster and make the sauce smoother.**

4. Stir in the cinnamon. Serve warm or let cool. Applesauce makes a yummy snack on its own, but it's also great over pancakes, granola, and in cake recipes (like the Applesauce Cake in a Mug recipe before this one). Store applesauce in an airtight container in the refrigerator, and it should keep for several weeks.

FIGURE 3-28: **After the apples are cooked, carefully remove the peels with a fork.**

MORE ABOUT COOKING FROM SCRATCH

Off the Shelf: Homemade Alternatives to the Condiments, Toppings, and Snacks You Love by Kris Bordessa (attainable-sustainable.net/off-the-shelf/)

The Omnivore's Dilemma: The Secrets behind What You Eat (Young Readers Edition; Dial Books, 2015) and *Food Rules: An Eater's Manual* (Penguin Books, 2009) by Michael Pollan

This Is What You Just Put in Your Mouth? From Eggnog to Beef Jerky, the Surprising Secrets by Patrick DiJusto (Three Rivers Press, 2015)

The New Bread Basket: How the New Crop of Grain Growers, Plant Breeders, Millers, Maltsters, Bakers, Brewers, and Local Food Activists Are Redefining Our Daily Loaf by Amy Halloran (Chelsea Green Publishing, 2015)

Grow Your Own Ingredients

4

No matter the weather, you can enjoy garden-fresh vegetables year-round!

FIGURE 4-1: Bring kitchen scraps back to life for quick and easy baby vegetables.

Fresh vegetables are tasty, crisp, and full of vitamins and minerals that keep you healthy. One of the best ways to guarantee a steady supply of fresh vegetables is to grow your own. But starting a vegetable garden from seed can be a lot of work. First, you need a patch of land that gets lots of sun. You may need to dig up the soil to loosen it and add in nutrients that plants need. If the weather's dry, you'll have to get the hose out. When weeds pop up, you have to go pluck them. And waiting until your vegetables are ready to harvest can take months.

Luckily, you don't need a garden to grow the vegetables in this chapter. And you can start to enjoy some of your indoor crops in just a few days! You don't even need any special equipment. Of course, you still have to make sure your plants get enough light and water, but most of these indoor gardening projects require very little work once you get them started. In fact, they'll probably inspire you to find more ways to grow your favorite veggies, indoors and out!

PROJECT:
GROW SPROUTS IN A JAR

Sprouts are newborn plants, complete with tiny roots, stems, and leaves. You harvest them just a few days after they "hatch" from the seed. To grow them, all you need to do is add some water, and let them do their thing. The seeds themselves contain everything the new plant needs for the first few days of life. And even though sprouts are small, they taste very much like the full-grown plant they come from.

Fresh sprouts are great in sandwiches and salads. If you like mild greens, start with alfalfa sprouts. If you're a fan of

MATERIALS

Quart-sized (1 L) canning jar or other sturdy, clear container

Small piece cheesecloth

Rubber band or rim from canning lid to screw onto jar

Seeds for sprouting, such as alfalfa, radish, or peas

Large bowl

Strainer

Optional: **Salad spinner**

FIGURE 4-2: **Alfalfa sprouts ready for a salad or sandwich**

spicy foods, try radish. Or if you really want to add crunch to your meal, try sprouting peas or beans. Be sure to use seeds that are specifically for sprouting. You can find them at your local health food store. You can also try online retailers like sproutpeople.org for sampler packs that let you test out small amounts of different kinds.

1. Go through your seeds and pick out any foreign objects like tiny pebbles.

2. Measure out 1 tablespoon (15 mL) of small seeds like alfalfa, or 3 tablespoons (45 mL) of larger seeds, like radish. Place them in your sprouting jar.

FIGURE 4-3: **A wide-mouth canning jar is handy for growing sprouts.**

3. Cut a piece of cheesecloth big enough to fit over the top of a quart jar with enough extra to hang over the sides. If your seeds are tiny, you can double the cloth. Cover the top of your jar with the cloth. Keep it in place by stretching a rubber band around the outer edge of the jar, or screwing on the rim from a canning lid.

4. Put the jar in a dark spot at room temperature where it won't be disturbed. Allow the seeds to soak for 8 to 12 hours.

FIGURE 4-4: **Use the ring from a canning jar lid to hold the cloth on.**

5. Leaving the cloth in place, pour the water out of the jar. Then fill the jar again part-way. Swirl the water around a little and drain the jar again. Stick the jar back in your shady spot. Rinse and drain the seeds the same way two or three times a day for three days.

6. On the fourth day, move the jar to a brighter location. Avoid direct sun—it can cook your sprouts. Continue to rinse and drain once or twice a day.

7. By day five or six, most of your sprouts will have green leaves. They are ready to be harvested—but first you should remove the *seed hulls*, the woody shells that covered the leaves. Transfer the sprouts to a bowl and fill it with cold water. With clean hands, gently pull the tangled-up ball of sprouts apart so the hulls separate from the plants. Move the sprouts to another container. Remove the hulls and repeat if needed.

8. When you've removed as many hulls as possible, put the sprouts in a strainer and let them drip dry for several hours. You can also use a salad spinner to remove the water.

9. When the sprouts are pretty dry, store them in the refrigerator in a zip-top bag or covered container. Your sprouts will stay crispy and delicious in the refrigerator for several days.

FIGURE 4-5: **It only takes a few days for sprouts to fill the jar.**

FIGURE 4-6: **Without light, the leaflets are yellow.**

FIGURE 4-7: **Rinse the sprouts to loosen the hulls.**

WHY DO PLANTS TURN GREEN IN SUNLIGHT?

Plants have a superpower: they can make their own food out of thin air (and a little water). The process is called *photosynthesis*, from the Greek words for *light* and *put together*. And it's the green stuff in plants that helps them do it. That green stuff is a chemical known as *chlorophyll*, and plants churn it out when they are exposed to sunlight (or artificial light, like that from a grow bulb).

Chlorophyll can absorb the energy from light and use it to power a plant's own chemistry laboratory. Inside the plant are cells that can take carbon dioxide gas from the air, along with hydrogen and oxygen (the elements that make up water), and recombine them into molecules of sugar. The sugar becomes food for the plant—and anyone who eats the plant. And as a handy byproduct, leftover molecules of oxygen are released back into the air (or water, if that's where the plant lives). It's a nifty trick, and it helps keep life ticking along on our planet.

PROJECT:
GROW INSTANT INDOOR VEGGIES FROM KITCHEN CUTTINGS

Sprouts are fun, but you don't need seeds to grow your own vegetables. Get a jump-start on harvesting an indoor salad by bringing your leftovers back to life. With only a dish and a little water, you can make your favorite green plants produce fresh tender baby leaves in no time!

Some zombie plants can live for weeks in just water. Others thrive even longer in a flower pot or container. Depending on the type of plant, you can keep picking new leaves for weeks or months.

FIGURE 4-8: **A garlic shoot peeking out from the clove.**

✦ GROW LiGHTS

Greens don't need as much light as vegetables like tomatoes or peppers, but it's still best to set them in a sunny window. If you don't have enough light, try an inexpensive grow light. They come in tubes like fluorescent lights, and can be mounted to the underside of a shelf or cabinet.

1. For scallions and leeks, cut off the bottom and set it in a cup of water, making sure the green leafy part is above the water line. For garlic, take one fat clove (the sections that make up the whole bulb) and set it in a cup of water, pointy side up. You should see new green shoots in a few days. Garlic shoots, called *scapes*, can be sliced up into short round tubes and sprinkled onto salads or other dishes, just like scallions. To harvest onion-y plants, just snip off what you need, right down to the base. New shoots should emerge so you can keep going for more cuttings.

FIGURE 4-9: **You only need the bottom of the leek to grow a new one.**

FIGURE 4-10: **Within a couple of days new green stalks emerge.**

2. For celery, romaine lettuce, and other vegetables where the leaves grow from a sturdy base at the root, cut off a good-sized chunk of the base (about 2 inches [5 cm] tall). Set it in a shallow cup or dish and add enough water to cover the bottom, making sure the green leafy part is above the water line. When you're ready to use the leaves, cut them down to the base. New leaves should keep growing for a few more cuttings.

FIGURE 4-11: **Romaine is a good lettuce to grow because its leaves are compact.**

FIGURE 4-12: **Tender baby leaves can be used for garnish or added to sauteed veggies.**

3. For basil and other herbs where the leaves grow off of thin stems, find the main stem or a branch where new leaves are forming at the end. Cut off a piece about 3–4 inches (10 cm) long, right below where a pair of leaves meet the stem or branch. Removing all but the top leaves will help new roots to grow along the stem (save these leaves to use in your cooking). Stand the plants up in a cup of water, letting the leaves rest on the rim if needed. Make sure the water covers a good portion of the stems. It will help to use a cup that blocks light from reaching the stem, since light can prevent roots from sprouting along the stem. To harvest your stemmy herb, pinch off a

FIGURE 4-13: **While other vegetable scraps may only produce a few leaves, a basil cutting can produce a whole new plant that will keep going for months.**

FIGURE 4-14: **If you pinch the stem right above a set of baby leaves you'll get new branches on your plant.**

stem or branch right above a pair of leaves with little baby leaves starting to show. Those baby leaves will then grow and branch out, producing new growth you can harvest.

4. You can regrow the green leaves on beets and carrots, although the main root will not grow back. Take a small root and cut off the top around the thickest part. Set it in a shallow dish and add enough water to keep the bottom covered. Harvest the beet greens by snipping them off whenever you want. Young and tender beet greens are good for spicing up a salad. When they're bigger you can saute them lightly in a frying pan with other veggies. Carrot tops resemble parsley and are good for throwing in soups and stews.

5. Give your plants as much light as possible without letting them get too hot. Make sure your plant is always in contact with water. Change the water as often as possible, at least every few days.

FIGURE 4-15: If your beet already comes with the leaves intact, cut them off and use them. The leaves will grow back.

FIGURE 4-16: You only need the top of the root.

FIGURE 4-17: Tender beet leaves are colorful as well as tasty.

6. For all plants, once you have some roots that look sturdy enough to survive being moved, you can plant them in a flower pot or other container. First, prepare the pot. If it has a large drain hole, cover the hole with a flat pebble or piece of broken flower pot so the soil doesn't wash away when you water it. Cover the bottom with loose pebbles for drainage. Figure out how far the roots of your plant will reach in the pot, and add potting soil to that level. Then take the cutting by the stem or gently by the leaves. Hold it over the pot so the roots dangle in. Gently sprinkle more potting soil around the roots until the plant can stand upright on its own. If there is a drain hole, put the pot in a saucer. Water the soil until it is completely moist. Do not over-water—there should not be puddles.

FIGURE 4-18: **Use a small flat stone to keep the soil from pouring out of the drainage hole.**

FIGURE 4-19: **A layer of gravel will help keep the soil from getting too wet.**

FIGURE 4-20: **Gently fill in around the plant with loose soil.**

FIGURE 4-21: **Potted plants like this garlic scape can last several months or longer.**

PROJECT:
PLANT AN (ALMOST) INSTANT AVOCADO TREE

FIGURE 4-22: **Avocado trees make nice houseplants.**

Growing any plant from a seed takes a long time. In the case of a fruit tree, you're usually talking years. But avocados are special. They produce giant seeds, and their seedlings are huge. Within a couple of months, you can have an avocado seedling that resembles a miniature tree. Keep it in a pot and you've got a nice indoor plant. If you take good care of your avocado tree, and you're lucky, you might even get it to bear fruit in a few years. If you live in a warm climate, you can transplant your avocado seedling into your yard and start your own avocado orchard.

1. Cut around the avocado down the middle the long way, being careful not to cut the pit inside. Twist the two halves and pull them apart. Scoop out the pit and wash it off. Save the edible part of the avocado

to make guacamole (see the following recipe), or slice it up and put in on a sandwich or in a salad.

2. Take the pit and hold it pointy side up. Stick toothpicks into the pit, an equal distance from each other around the middle. Set the pit in the glass, letting the toothpicks rest on the rim to hold the pit up. Add enough water to reach the bottom of the pit. Change the water every day or two, and make sure it always covers the bottom of the pit.

FIGURE 4-24: **Carefully cut around the pit.**

3. In a few weeks, the seed will appear to crack open, and a long root and a stem should appear. (Be patient; it may take as long as two months. If nothing has happened by then, try again with another pit.) When the stem is about 6 inches (15 cm) long, cut it down to about half its size. Wait until the stem has sprouted new leaves, then plant the pit in a large pot. Make sure to let the top half of the pit stick out of the soil.

4. Place the pot in a sunny spot, and keep the soil moist. If you want your tree to develop

FIGURE 4-25: **Place the pit, fat end down, in the water.**

branches, cut the stem in half when it is about 1 foot (30 cm) long. This will make your miniature indoor avocado plant more tree-like.

RECiPE: GUACAMOLE DiP WiTH TOMATO

FIGURE 4-26: **Use your avocado to make spicy guacamole dip.**

1. Put the avocado in a small bowl and mash it with a fork until mostly smooth and creamy.

2. Mix in tomato pieces.

3. Squeeze in some lime juice, and mix in chili pepper and salt.

4. Serve with tortilla chips or solar nachos (see Chapter 5 for recipe).

FIGURE 4-27: **Avocado mashes easily with just a fork.**

FIGURE 4-28: **Tomatoes should be cut into small cubes.**

PROJECT:
MAKE AN AQUAPONIC JAR

MATERIALS

Large, wide-mouth canning jar or sturdy recycled plastic jar, ½ gallon (2 L) or larger

Recycled cup or container no larger than the jar opening

Wide stiff straw (if cup sits right in the jar opening)

Lid or foam plate to cover jar if cup needs support

Pea-sized gravel from a garden shop (rinse well with plain water) or aquarium gravel

Aquarium plant for oxygenation

Zebra danios, minnows, or other small hardy fish—no more than 3 per jar to start, or 1 very small goldfish if your jar is 1 gallon (4 L) or larger

Goldfish flakes

Extra-long, round cocktail toothpick

Leafy plant started from cutting, such as basil, mint, or lettuce (see the "Instant Indoor Veggies" project earlier in this chapter), or hydroponic plant from grocery (just transfer it right into your planter)

Stiff cardboard to make shades to block direct sunlight from hitting the water

Do you like pets and fresh vegetables? With an aquaponic jar, you can have both! *Aquaponics* is an offshoot of *hydroponics*, a method of taking plants that normally live in soil and growing them in water instead. Hydroponics is a very efficient way to grow food, but it takes a lot of maintenance. Big indoor hydroponic farms that grow tomatoes and herbs use computerized systems to make sure the plants get what they need as soon as they need it. The systems constantly monitor the plants to control

FIGURE 4-29: **An aquaponic jar is a complete ecosystem—no filter needed.**

patterns of light and dark and add chemical fertilizers to the water in exact amounts.

In aquaponics, plants grow in an aquarium—and the fish provide all the fertilizer the plants need. An aquaponic jar is a complete ecosystem. That means that different living things in the system help each other. When an aquaponic system is in balance, every part helps keep it healthy. This is how it works:

* You feed the fish. (You also make sure the aquarium has clean water, sunlight [or artificial light], and doesn't get too hot or too cold).

* The fish eat the food and excrete waste (fish poop).

* Microscopic bacteria growing on gravel on the bottom of the fish jar digest the fish poop and break it down into nutrients, like nitrogen, that help plants grow.

* If you have snails in your aquarium, they will help keep it clean by eating algae as they crawl along. Algae is a tiny green plant that can cover the walls of the jar with slime.

* Plants in the aquarium absorb the nutrients and produce oxygen that the fish need to survive.

* Veggies growing on top of the aquarium also absorb the nutrients when they in water through their roots.

* You eat the veggies!

Everything you need to make an aquaponic jar can be found in your local food store, home and garden shop, and pet store. The jar itself can be a large canning jar or kitchen storage jar. You can even recycle a sturdy, oversized, clear plastic jar from the grocery—the kind that pretzels or other snacks sometimes come in. Just remove any labels and wash it well before using. The planter for your veggies can be made from a plastic cup or recycled yogurt container.

As for the fish, you can find small hardy species like zebra danios, minnows, or goldfish inexpensively at most pet shops, as well as aquarium plants like hornwort. When picking out an aquarium plant, ask the store clerk to include any little snails that might be attached. They will usually throw them in for free.

⤳ AQUAPONIC SAFETY RULES

Make sure anything you put in your aquaponic setup is safe for fish and plants (and you!). You don't want anything, such as decorations, in the tank that is made of a material that could break down and release harmful chemicals, such as certain metals. Also, avoid rocks and materials that might change the pH of the water, such as high-pH limestone. Rinse off all decorations, rocks, gravel, and other material before you place them in the water to remove dust that can make the water cloudy. And always wash off the vegetables you are growing in your aquaponics system before you eat them—just as you should do with all produce!

1. First, check that the recycled cup you are using for your planter—or as it's known in hydroponics, the *net pot*—isn't too big or too small. When the top of the cup is slightly above the rim of the jar, the bottom of the cup should be a little ways into the water. Carefully cut slits in the sides of the cup wide enough to let water in without letting the gravel fall out. Poke a few holes in the bottom of the cup from the inside with a long nail or other long sharp object.

FIGURE 4-30: **The straw should be long enough to stick out both the top and bottom of the cup.**

If your cup fills the opening of the jar, you will need to insert a straw through the bottom of the cup to let in air. Poke a hole right next to the side of the cup. Widen the hole carefully by working a pencil into it until it is big enough for the straw to fit through snugly. Trim the straw a little above and below the cup. Make sure you have a toothpick or bamboo skewer longer than the straw for feeding the fish (more about this in Step 7).

2. If the planter for your veggie is too small to sit on the rim of the jar by itself, make a holder for it using the lid of the jar or a stiff foam plate.

Trace around the cup and cut a hole in the lid or plate that's a little smaller than the rim of the cup. Make sure that the lid or plate is sturdy enough that it won't sag when the planter is full of gravel. Poke two or three small holes in the lid to let air circulate and let you feed the fish.

3. Next, put together your fish jar. Fill a pitcher with about as much water as your jar can hold, and let it sit until it reaches room temperature— this is important because extreme temperature changes are not good for the fish. This will also give any chlorine in the water time to evaporate. Meanwhile, pour 1–2 inches (2.5–5 cm) of gravel in the bottom of the jar. You can also put in decorations like small, clean stones or tiny, clay flower pots to give your fish some hiding places. Then, add the water. Leave some airspace at the top of the jar. Put in the aquarium plant (and snails, if they came along). Finally, insert the planter. Give the jar a day or two to allow helpful bacteria to start to grow on the gravel.

4. While you're waiting, plant your veggie in the cup. Cover the bottom of the cup loosely

FIGURE 4-31: **Cut the hole in the lid carefully so it doesn't tear and let the net pot slip through.**

FIGURE 4-32: **The aquarium gravel provides a lot of surfaces where the good bacteria can grow.**

FIGURE 4-33: **Snails help keep the jar clean by eating algae.**

FIGURE 4-34: **Plant some basil cuttings with roots in the cup.**

FIGURE 4-35: **Carefully fill in around them with gravel to anchor them in place.**

with gravel. Insert the plant into the cup so the leaves are above the rim. Gently add more gravel around the plant to help hold it upright. Then carefully set the cup in the jar (or the holder on top of the jar).

5. Time to move your fish to their new home. Remove the cup (or the entire holder with the plant cup still in place). Set it in a temporary holder, such as another jar tall enough to hold the planter without damaging any roots sticking out underneath. Gently transfer the fish to the jar, being careful to avoid stirring up the water any more than is necessary. Then replace the planter. Your aquaponic fish jar is complete!

6. Keep your aquaponic system in a sunny spot where your veggies can get as much light as possible. However, it's important to avoid too much light in

FIGURE 4-36: **The yogurt cup sits snugly in the top of the canning jar.**

the jar itself, which can speed the growth of algae and make the water uncomfortably warm. If light shining directly on the jar is a problem, make a shade for it. Cut a rectangle of stiff cardboard as high as the jar, and wide enough to fold around the jar on three sides. Stand it up on sunny days.

7. Use a long toothpick to feed your fish without disturbing the plant. First, crush up all the goldfish flakes into tiny pieces. (If your food canister is large, transfer some to a smaller container with a top that lifts off, like a pill bottle. Refill the smaller container as needed.) Then, take the toothpick and wet the tip by dipping it in the fish jar. You should be able to poke it through a hole in the holder or insert it into the straw. Pull it back out and use the damp end to pick up some of the crushed flakes. Now dip the flake-covered toothpick back in the water and shake it around until the food floats off. Do not overfeed your fish! Each fish only needs a few small flake pieces per day. It's OK to skip a day or two if you're away.

FIGURE 4-37: **The shade can be any color, or even show an underwater scene.**

FIGURE 4-38: **Insert the toothpick with its tip covered in food into the straw, and poke it down just far enough so that the food floats off into the water.**

8. Although you don't need to monitor your aquaponic system on a constant basis, you will have to do some maintenance to keep everything in balance. First, make sure the hole or straw for the toothpick doesn't become clogged with dried food. Watch the water level, and add more when it gets below the level of the planter. If the roots or underwater plant start to take up

too much space in the jar and interfere with the fish, carefully pull them out and trim them with scissors.

9. Once every few weeks, or as needed, replace some of the water in the jar with fresh water. Fill a container with tap water and let it sit, as in Step 1. With a paper cup or a scoop, remove about half the water. Be careful not to scoop up the fish! Then, slowly pour in fresh water, being careful not to stir up debris that will make the water cloudy. Next, replace the planter. If the jar starts to become covered with slimy green algae or brown bacteria, it needs a more thorough cleaning. Use a small cup or fish net to move the fish to a separate temporary container. Empty out most of the water from the jar, and wipe down the inside with a paper towel. Do not use soap! Be sure to wipe off any stones or other decorations that need it, as well.

FIGURE 4-39: **A well-balanced aquaponics system**

10. Harvest your greens, as you did in the "Grow Instant Indoor Veggies from Kitchen Cuttings" project. Replace the plants, as needed, and your aquaponic system can keep going for a long time.

PROJECT:
USE A COMPOST JAR TO TURN VEGGIE SCRAPS INTO SOIL

One of the nice things about growing food plants is that nothing goes to waste. You can even use vegetable scraps to make rich soil to help new plants grow by composting. *Composting* lets tiny microorganisms break down the old plant material into minerals and other nutrients. There are all sorts of fancy compost bins you can build, but really, all you need is a spot in your yard where you can start a small pile. Or if you don't have space outdoors, go on to the next project to learn how to build an indoor worm bin and bring the critters to you!

1. Save any kind of raw vegetable scraps (but not animal products like dairy, eggs, or meat) in the jar on your counter. Keep it closed so it doesn't attract flies.

2. When the jar is full, dump its contents onto your pile. Cover the fresh scraps with older material.

3. Every few days, use a shovel or rake to turn over the materials in the compost pile and mix them up. Within a few weeks, bacteria, fungi, and tiny critters like protozoa and nematodes should eat the scraps and turn them into rich black compost. Dig up the compost and sprinkle it around as fertilizer on the plants in your garden or indoor pots.

FIGURE 4-40: **A clear compost jar is colorful, but if you prefer, you can also use a plastic, mini trash can or other lightweight washable container.**

PROJECT:
BUILD AN INDOOR WORM BIN

Worm bins aren't as icky as they sound. In fact, once you've set one up, you'll hardly notice it's there. The trickiest part is finding the worms. The best kind of worms for an indoor bin are not the earthworms you see on the sidewalk after a rain but red wigglers (scientific name: *Eisenia fetida*). They like to stay near the surface, so they do fine in shallow boxes. Red wigglers grow to be about 1½–4 inches (2.5–10 cm) long, and eat more than their own weight in food every day. They will also reproduce rapidly, and lay up to 20 eggs a week. You may be able to find red wigglers by digging in a compost pile or under a pile of decaying leaves. They are also sold as bait, commonly called "trout worms."

MATERIALS

Medium to large plastic box with lid (dark colored is best, since worms don't like light)

Newspaper or other scrap paper (not color or glossy pages), torn into strips about 1 inch (2.5 cm) wide and 6 inches (15 cm) long

Vegetable scraps

Small container of worms

FIGURE 4-41: **A sweater box with holes punched in the lid makes a compact home for worms.**

FIGURE 4-42: **As soon as you move the worms into the bin, they'll start to burrow down to the lower layers.**

🥕 WHAT KEEPS THE WORMS FROM ESCAPING?

A worm bin can be kept under the kitchen sink or other cool, dark place in your home, and the worms will be perfectly happy. Although you need air holes to let the worms breathe, they like to stay buried, so there's no reason for them to try to make a break for it. If you do get a runaway, just pick it up gently and put it back in the bin. They don't move very fast. They're worms.

1. Clean your box with soap and water and remove any labels. Make air-holes in the top using a nail and hammer.

2. To make bedding for your worms, start by filling the box halfway with the newspaper strips. Wet the paper by sprinkling or spraying water on it. The newspaper should be damp, but not dripping.

3. Gently add worms, along with any dirt they are in. Cover with a layer of damp paper strips.

4. Add a small amount of food scraps, and cover with another layer of damp paper strips. You can feed your worms any kind of raw fruit and vegetable scraps, including tea bags (without the staples) and coffee grounds and filters. Rinsed egg shells are OK and provide calcium, but other animal products will attract flies.

FIGURE 4-43: **Worms can digest newspaper, so it makes good bedding for a worm bin.**

5. Cover the box and keep it in a cool, dark, quiet place, such as under the kitchen sink.

6. Remember to check on your worms every few days and add some food scraps if needed. To add food, make a small space in the bedding, dump in your

FIGURE 4-44: **Whenever you add new food scraps, cover them with some of the paper.**

food scraps, and then cover with some of the bedding. Add just a small amount of food at first. You can increase the amount of food as baby worms add to the population.

7. The worms will eat both the food and the newspaper bedding. The worm poop they leave behind is known as castings—this is the fertilizer! To make it easier to harvest the castings, encourage the worms to move to one side of the box by only leaving food on that side. Then, scoop out the poop on the other side. You can put castings on your house plants or garden for fertilizer.

MORE ABOUT GROWING EDIBLE PLANTS

Sprout People (sproutpeople.org)

KidsGardening.org (kidsgardening.org)

Do the Rot Thing: A Teacher's Guide to Compost Activities (www.cvswmd.org/uploads/6/1/2/6/6126179/do_the_rot_thing_cvswmd1.pdf)

The Aquaponic Source (theaquaponicsource.com)

AquaJar aquaponics jar kit (http://myaquajar.com/en/)

Home Ecology aquaponics kit for fish tanks (http://www.homeecology.net/)

Don't Throw It, Grow It! 68 Windowsill Plants from Kitchen Scraps by Deborah Peterson and Millicent Selsam (Storey Publishing, 2008)

5

Cook off the Grid

Here are some ideas for cooking a meal without spending a lot of time in the kitchen.

FIGURE 5-1: Hot, crispy cookies baked in a cardboard oven that's powered by the sun.

If you've ever roasted marshmallows over a campfire, grilled hot dogs, or made spaghetti on a camping stove, you know that everything seems to taste better when you cook it outdoors. But wood, charcoal, and gas aren't the only fuels you can use outdoors. You can cook "off the grid" using homemade fuel, renewable energy, or sometimes no extra power at all.

Like barbecue grills and campfires, these cooking methods are great for picnics, backpacking, and other fun activities away from home. But they also serve a more serious purpose. In emergencies, you can prepare food on an alternative cooker even if there's no power. And in parts of the world where the only way to cook is to burn firewood, dried manure, or other fuel, alternative cookers can save hours of labor and cut down on smoke in the air that makes it hard to breathe.

Some of the alternative cooking methods in this chapter have been around for thousands of years. Others are more recent, with new and better versions being invented all the time. Try these alternative cooker ideas and see what improvements you can come up with!

⇒⟶ TEMPERATURE AND TIME

Many of your favorite recipes probably use a standard oven temperature of 350°F (180°C). But some alternative cooking methods don't get as hot as your stove or your oven, or keep a steady temperature throughout the cooking process. You may have to cook a dish longer to get the same result—for instance, a recipe that takes 30 minutes in your kitchen might several hours in a homemade cooker.

Changing the temperature and the cooking time can also change the way the dish looks, or give it a different texture. Cakes and other baked goods are usually moister than in a traditional oven.

FIGURE 5-2: **Toasty warm in the solar oven.**

How can you tell whether a recipe will work in your alternative cooker? It depends on how temperature affects the ingredients in the recipe. Cooking is chemistry, and sometimes you have to reach a certain temperature to set off chemical reactions that create the results you're looking for. Use this chart to help you predict how a recipe may turn out. You can also search for versions designed for slow cookers or other low-temperature alternatives.

TEMPERATURE	INGREDIENT
90°F (35°C)	Fats like chocolate and butter start to melt and release moisture, making baked goods puff up.
150°F (65°C)	Eggs begin to set and become solid.
160°F (70°C)	Vegetables start to get soft. Dishes that are supposed to stay moist, like rice, can be simmered at a slow boil.
180°F (82°C)	This is the lowest temperature for baked goods like cakes and cookies.
212°F (100°C)	A pot of water reaches a full boil, hot enough to cook pasta or steam vegetables.
300°F (150°C)	The Maillard reaction (a chemical change that happens when amino acids combine with sugar) starts to make meat turn brown and smell yummy. Sugar starts to caramelize (turn dark brown and crunchy).

Tip: When using alternative cookers, avoid peeking! Every time you open your cooker, you let out heat, which can slow down the cooking process even more.

 ## TEMPERATURE AND FOOD SAFETY

When cooking anything you wouldn't eat raw—especially meat and eggs—be sure your cooker reaches a temperature of at least 180°F (82°C)! That will kill off any harmful bacteria that may be lurking there. Use a thermometer that doesn't require you to open the oven to check the temperature. A probe thermometer can be poked right through the cardboard wall of your solar oven.

Also, be sure to use oven mitts or a hot pad when taking your food out of your solar oven. It gets hot in there!

COOKiNG WiTH THE SUN

The sun is a great big ball of energy in the sky—and a solar oven is an alternative cooker that turns the energy in sunlight directly into heat. Just like a car with its windows closed on a bright, clear day, a solar oven heats up just by sitting there. Under the right conditions, the best solar ovens can get as hot as a regular oven.

The first known solar oven was built in 1767 by a Swiss physicist and mountain climber named Horace-Bénédict de Saussure. Saussure's "hot box" looked like a miniature greenhouse. It consisted of five glass boxes, one inside the other. When he put it in the sun on a black table, the temperature inside climbed to 230°F—hot enough to bake fruit. Saussure even carried it with him to the top of a mountain in the Swiss Alps. Even though the air was much colder up there, the sun was just as bright, and the box still reached the same temperature. That proved it was the sun that caused the box to heat up.

Modern solar ovens were first developed in the 1950s to help people in desert regions where it is hard to find firewood for cooking, but the idea did not catch on. In 1954 in sunny Phoenix, Arizona, researchers from around the world gathered to share ideas about practical uses of solar energy, including cookers. A scientist from the Massachusetts Institute of Technology (MIT) named Mária Telkes demonstrated a wooden box cooker with a double layer of glass on the top and four large reflectors around the sides that looks very much like standard solar cookers today. Also at the conference was the Umbroiler designed by George Löf

FIGURE 5-3: **Keep your solar oven pointed toward the sun.**

from the University of Denver in Colorado, a curved reflective cooker that could fold closed like an umbrella. Löf sold Umbroilers for a few years, but they were not a success—too expensive and too flimsy to hold up.

By the 1970s, solar cooking was mainly a hobby for people interested in alternative energy. Today, thanks to simpler and more efficient designs, solar ovens are again attracting attention as a way to help poor people in hot, dry areas that get lots of sun.

HOW SOLAR OVENS WORK

Solar ovens work because light and heat are just different forms of the same kind of energy. When light hits a substance, some of it bounces back in the form of light waves we can see, and some is absorbed. The absorbed light energy causes the molecules in the substance to jump around and heat up. I recommend using dark colors inside a solar oven because dark substances absorb more light than they reflect. That means dark-colored substances heat up more quickly than light-colored substances.

Solar ovens come in many different designs. A parabolic solar cooker like the Umbroiler looks like a large reflective bowl. Sunlight hits the bowl and bounces back toward one spot in the middle. A cooking pot placed in the middle of the bowl can heat up quickly. Box-type solar ovens like the one made by Telkes have a clear window that lets sunlight in. The window blocks heat waves and hot air from escaping. They don't get as hot as reflective solar ovens, so food cooks slower, but there's less chance of burning it. The box also protects the pot or pots inside from the wind, which helps keeps them at a steady temperature while cooking.

You can buy commercially made solar ovens in many different styles, but many people still like to design and build their own. Homemade solar ovens can cook almost as well as the commercial models.

PROJECT: BUILD A CARDBOARD SOLAR OVEN

MATERIALS

Large, flat, rectangular, corrugated, cardboard box with the opening on the narrower side, or with a pizza-box kind of lid

Similar box small enough to fit inside the large box with about 2 inches of space all around

Scissors able to cut cardboard boxes or a box cutter

String, brass brads, and a large button or other weight *or* about 8 feet (2.5 m) peel-and-stick Velcro tape

White glue

Aluminum foil (heavy duty is best)

Optional: Aluminum-foil duct tape (not regular duct tape, which gets soft in the heat)

Black construction paper (enough to line the bottom of the smaller box)

Extra corrugated cardboard

Masking tape

Oven-proof plastic roasting bag (turkey size is best)

Sheet of window or picture-frame glass, the same size or slightly smaller than the smaller box

Clean, recycled, shredded copy paper or recycled newspaper

Optional: Protractor

The classic cardboard box solar oven isn't as sturdy or efficient as commercial models, but it still can do a pretty good job. Use these directions as a rough guide—the exact design of your solar oven will depend on the cardboard boxes and other materials you have on hand. But to be efficient, any box-type solar oven should have these things:

* An outside container that's lightweight, sturdy, and easy to move around so you can turn it to keep facing the sun as it moves across the sky

* A big window to let in as much light as possible

* One or more reflectors to bounce additional sunlight toward the cooking pot

* Insulated walls that keep the heat from escaping

* A black lining to absorb light and turn it into heat

⇛⟶ SOLAR COOKING PANS

For best results, use dark-colored, thin-walled cooking pans. Look for them at dollar stores, thrift shops, or garage sales. You can make a covered baking pan that is small enough to fit in a solar oven by taking two small, dark-colored saucepans and removing the handles. Flip one upside down and place it on top of the other as a lid. Clamp them together with binder clips.

FIGURE 5-5: **Make a covered solar pan from two small saucepans.**

⇛⟶ CARDBOARD BOXES

The solar oven pictured here uses an outside box 20 inches long by 17 inches wide by 8 inches high. The inside box is 17 inches long by 11 inches wide by 7 inches deep. If you can't find cardboard boxes that are the size or shape you'd like to use, cut and reassemble a box to fit, using additional pieces of corrugated cardboard if needed.

FIGURE 5-5: **You need two flat, rectangular boxes to make an insulated box cooker.**

1. Make a flap in the larger box that you can lift up to become a reflector. To do this, tape the box closed and lay it flat. Center the smaller box on top and trace around it. Cut only along the sides and front of the shape you just traced. Make an indented fold line by tracing over the fourth line, pressing hard with a rounded pen or pencil point. Fold the flap back along the line.

2. To use string to hold the lid up, punch a hole with the tip of the scissors or a ballpoint pen through one of the top corners of the lid, and insert a brass brad loosely into the back. Punch another hole through the back of the box. Tie a sturdy piece of string onto the top brad. Tie a button or other weight onto the end of the string. When cooking, wrap the string around the bottom brad, as needed, to hold the lid up at the right angle to reflect sunlight into the box. As an alternative, attach strips of peel-and-stick Velcro tape on the lid and the box and connect them to hold the box open. (See Figure 5-20).

3. Cover the inside of the larger box, including the reflector flap, with aluminum foil. First, brush on, or use a craft stick

FIGURE 5-6: **Trace around the smaller box to mark where the opening on the larger box will be cut.**

FIGURE 5-7: **Use a straightedge to make a nice straight fold.**

FIGURE 5-8: **A string with a button tied at the end can be wrapped around a brad to hold the lid open at the desired angle.**

to spread on, a thin layer of glue. Then, press pieces of foil into place, smoothing as much as possible. Secure with foil tape, if needed.

4. If the smaller box does not already have a lid, cut one out the same way you did with the larger box. Make the lid a little larger than you want your window to be. Add an extra strip of cardboard along the front of the lid. Trim it to make a tab that will slide into the gap between the two boxes when they are assembled and hold the lid closed, as shown in Figure 5-14. Then, cut an opening inside the lid for the window, leaving about an inch all around as a frame. Line the inside of the smaller box with black paper or cardboard following the same directions as in Step 3.

5. Use scrap cardboard to make a "frame" the same size as the lid, with the same size opening. For the window, fold a closed roasting bag so it is a little bigger than your opening. Tape it over the opening in the separate frame you just made. Then, line up the frame with the lid so that the plastic is sandwiched in between. Attach it securely with more tape. If you have a piece of glass, you can use it instead of the plastic bag, or tape the glass on the outside of the window to help insulate it.

FIGURE 5-9: **Line the larger box with foil.**

FIGURE 5-10: **Make a window by stretching a plastic roasting bag over a cardboard frame.**

FIGURE 5-11: **The lid of the inside box with the black paper lining and window in place**

6. You will need to make stands or spacers to raise the inner box so its top is even with the top of the outer box. First, measure the difference between the height of the inner and outer boxes. Cut four strips of cardboard as wide as that measurement and about 9 inches long. Bend each strip in two places, and tape the top and bottom together to make a hollow triangle. Place the stands inside the larger box. Add enough shredded or wadded-up paper to loosely fill up the space that will be left between the two boxes. Insert the smaller box into the opening and tape it into place. Do not tape the front edge of the smaller box to leave an opening for the tab that holds the lid closed.

7. Your solar oven is now ready to use! Place it in a sunny spot that gets little or no shade during the day and point the window toward the sun. Place your cooking pot inside the cooker and close the window tightly so no air escapes. Use the string or Velcro strip to pull the reflecting lid back just enough so that it bounces sunlight into the cooking area. Put your oven out early—by

FIGURE 5-12: **Spacers and insulation help keep heat from escaping from the inside box.**

FIGURE 5-13: **Extra insulation will keep heat in.**

FIGURE 5-14: **The inner box is inside the outer box. Extra cardboard added to the front of the window lid has been trimmed to make a tab that tucks into the space between the two boxes and holds the lid down when cooking.**

late morning at the latest—to get the most sunlight. Check it at least once every half hour or hour to make sure the sunlight is still hitting the window as the sun moves across the sky.

8. In areas where the sun sits low in the sky, you can make your oven work even better by building a stand to hold it at an angle. Take a long, wide strip of cardboard and fold it into two triangles, as shown in Figure 5-16. Use a *protractor*—a math tool that lets you measure the angles between two sides of a triangle—to make sure the angles that touch are both 45 degrees. Tape or glue the strip securely. Build a second stand to hold the cooking pan level when the box is tilted, so your food doesn't spill. Take a wide strip of black cardboard (or cover a strip of cardboard with black paper) and fold it into a triangle with two 45-degree angles. Remember to check that your pan fits when the box is tilted.

FIGURE 5-15: **Make a tilt stand to hold your box at an angle. Another stand inside keeps your cooking pot level.**

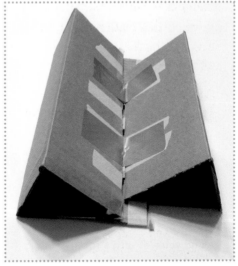

FIGURE 5-16: **A tilt stand to hold the solar cooker at an angle, facing the sun**

FIGURE 5-17: The inside angles of the triangles should measure 45 degrees on the protractor.

FIGURE 5-18: A smaller stand holds the pot level inside the cooker when it is tilted.

WHY 45 DEGREES?

Really, it's just to make the math simpler. It's easy to measure 45 degrees, and the space in between is 90 degrees, so the box will fit in it.

IMPROVEMENTS, ADDITIONS, AND VARIATIONS

- Make additional reflectors with cardboard and foil, and attach them to the outer box around the window. This will direct even more sunlight into your oven. If you attach them with Velcro, you can remove them to store the solar oven flat.

- For a more permanent solar oven, use kraft paper tape (the kind you wet with a sponge) or white glue instead of masking tape. For adjustable parts like the lid and the reflectors, use peel-and-stick Velcro tape.

FIGURE 5-19: Attach Velcro straps firmly to the outside of the box.

- Insert a probe-style oven thermometer through the walls of the cooker to measure the temperature in the inner box. Make sure to get an oven thermometer, not a meat thermometer. A thermometer with a probe long enough to reach the food will give you the most accurate reading. Or try an infrared thermometer that lets you take the temperature of the food from the outside using reflected light.

- Place a black metal tray (like a broiler drip pan or cookie sheet) in your oven as a "thermal mass." It will absorb heat and release it slowly. That way, you can let your oven preheat while you are getting your food ready, and it will keep some heat from escaping when you open the oven door. Or use a medium or large dark-colored metal cake pan as your inner "box."

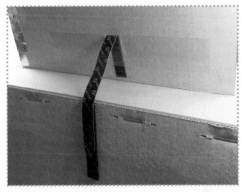

FIGURE 5-20: **This Velcro strap holds the lid up.**

- Put your cooking pan inside a second closed roasting bag to make an "oven within an oven." This will help retain heat around the cooking pan, and keep moisture from escaping and fogging up the oven window.

- Make your solar oven weather-resistant and more airtight by painting it or covering the outside with fabric. You can also seal with Mod Podge, watered-down glue, or shellac.

- Use a rolling cart to hold your solar cooker so it can be turned easily as the sun moves across the sky.

- Think of ways to make a sun guide to show you when your cooker is lined up exactly with the sun. For instance, you could add a sundial that points to noon when the sun is directly in front of it. Or add electronics and attach a light sensor to trigger an alarm (or send you a text) when it's time to move the box.

SOLAR RECIPES

The temperature that a solar oven can reach varies, depending on outside conditions (like sunshine, wind, and air temperature) and how well it is designed. But even under less-than-perfect conditions, you can still test out your solar oven with the following recipes. While higher temperatures will produce the best results, the recipes in this section are safe to eat, even if they're a little undercooked. Your baked goods may come out slightly gooey, but melted cheese or chocolate should be perfect. Mmmm . . .

RECiPE: SOLAR NACHOS

1. Use a dark plate or oven tray to hold your chips. You can put a sheet of aluminum foil on it to keep the cheese from sticking, but try to leave as much of the dark-colored tray showing as possible.

2. Arrange a heap of chips on the plate, but not too high. Sprinkle shredded cheese on top.

Check your nachos in 30 minutes—sooner if it's a hot day. When the cheese is melted to your taste, remove and enjoy! Serve with guacamole (see Chapter 4).

MATERIALS

Tortilla chips (blue chips work great)

Shredded Mexican cheese (or cheddar, jack, or your favorite)

FIGURE 5-21: **Nachos cook relatively quickly in the solar oven.**

RECIPE: SOLAR OATMEAL COOKIES

🥄 MATERIALS

1 cup oats

1 cup flour

1 pinch salt

$\frac{1}{3}$ cup oil

1 teaspoon vanilla extract

$\frac{1}{2}$ cup brown sugar or honey

$\frac{1}{2}$ teaspoon baking soda

$\frac{1}{2}$ teaspoon baking powder

$\frac{1}{2}$ cup chocolate chips or raisins (or both!)

$\frac{1}{2}$ cup water

1. Mix ingredients together to form a sticky dough.

2. Drop by the spoonful onto a cookie sheet or a mini muffin pan.

3. Place the pan in the oven. Watch to see when the cookies expand and crisp up. It may take 30 minutes to several hours, depending on conditions.

Tip To speed up cooking, make a cover for the cookie sheet with an overturned dark pan. Or add a teaspoon of cocoa to make the dough darker.

FIGURE 5-22: Oatmeal cookies come out chewy and delicious.

FIGURE 5-23: Ready for the oven. Leave space around each one; they spread.

RECIPE:
SOLAR CHOCOLATE CAKE

MATERIALS

Dry ingredients:

$\frac{1}{2}$ cup flour

3 tablespoons unsweetened cocoa powder

$\frac{1}{3}$ cup sugar

$\frac{1}{4}$ rounded teaspoon baking soda

Pinch salt

Wet ingredients:

$\frac{1}{3}$ cup cold water

2 tablespoons oil

1 teaspoon vinegar

$\frac{1}{2}$ teaspoon vanilla

$\frac{1}{2}$ cup chocolate chips

Spray oil (or oil or softened butter)

Toothpick

FIGURE 5-24: A small, solar chocolate cake makes a nice snack.

1. Use a wire strainer to sift the dry ingredients together in a medium bowl.

2. Add the wet ingredients to the bowl. Mix well. Fold in the chocolate chips.

3. Spray the inside of the baking pan with oil (or rub on oil or butter with a paper towel or brush). Then, sprinkle in a spoonful of flour or cocoa and shake the pan around until the bottom and sides are coated evenly.

FIGURE 5-25: **Chocolate cake batter in the covered pan. The inside has been sprayed with oil and coated with flour to prevent sticking.**

4. Pour the batter into the pan and cover it. Place the pan in the oven for one to three hours, depending on how hot your oven gets. To test that it is done, stick a toothpick into the middle. If it comes out clean, your cake is ready! Because of the chocolate chips, it may still be a little gooey. You can eat it warm from the oven, like a lava cake, or let it cool.

SELF-CONTAINED COOKING

Many busy families swear by slow cookers. Just throw in your ingredients, turn it on to a low or medium temperature, wait a few hours, and ta-da! A nice hot meal, no sweat.

Thermal cookers work the same way, but instead of an electric heater, they rely on insulation to keep the food at the proper temperature. First, the food is heated on the stove. Then it's popped into the thermal cooker, where it continues to cook itself until it's time for dinner. Early thermal cookers included Native American cooking baskets that were lined with clay and filled with hot stones. Some traditional cultures put the cooking pot right in a hole in the ground to keep it warm. And European immigrants to the United States brought with them the hay box, which used hay or straw for insulation.

In 2008, Sarah Collins and Moshy Mathe drew on these ancient ideas and created the Wonderbag to help poor women in their country, South Africa. The Wonderbag looks like a big, stuffed pillow with a hole to fit a cooking pot and a padded lid. Collins first got the idea during a power outage, when she remembered her grandmother wrapping a pot in blankets to keep it warm and let it finish cooking. These projects use materials you have on hand—such as baskets, lunch bags, and blankets—to create soft thermal cookers you can try at home.

PROJECT: SELF-CONTAINED COUNTERTOP YOGURT-MAKER

MATERIALS

1-quart (1 L) size heat-proof container with a tight lid, such as a travel mug

Candy thermometer

Medium saucepan

Large saucepan or bowl

$3\frac{3}{4}$ cups (800 mL) milk (slightly less than the size of the container)

1–2 tablespoons fresh yogurt with live cultures (or 3–4 crushed probiotic pills)

Insulated lunch bag or heavy towels

Yogurt is another fermented food. It's made by adding friendly bacteria to milk and keeping the mixture warm for several hours. Heating the milk first changes the chemical structure and helps it thicken. The bacteria help,

FIGURE 5-26: **Make homebrew yogurt right in the container.**

too, by turning the milk slightly acidic and sour. You'll start with live cultures from store-bought yogurt or probiotic pills (which help people who need to replace the bacteria that normally live in the human stomach). It's easy to put together a yogurt-maker using a jumbo travel mug and an insulated lunch bag. A one-quart (one-liter) container will make enough yogurt for several servings.

1. Put the candy thermometer in the medium saucepan, clipped to the side so it doesn't bang around. Fill the larger saucepan with cold water. Set out a clean, insulated mug (or similar-sized container with a tight lid). Also, set out an insulated lunch bag or heavy dish towels.

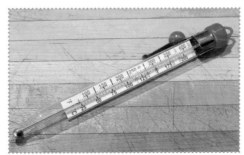

FIGURE 5-27: **A candy thermometer will help you keep the yogurt in the right temperature range.**

2. Let the fresh yogurt starter sit out while you heat the milk. Pour the milk in the smaller saucepan and set it over a medium flame, watching it constantly to make sure it doesn't boil over. When the milk reaches 180°F (82°C), it should just be starting to bubble. Take the saucepan off the stove and set it in the larger pan of cold water to cool.

FIGURE 5-28: **Watch the pan as the milk heats to make sure it doesn't start to boil.**

3. When the milk has cooled to 125°F (52°C), take it out of the cold water. Remove the thermometer. Add the starter yogurt and stir well.

4. Carefully pour the milk into the container and close it tight.

FIGURE 5-29: **Cool it in a pan of cold water.**

Place the container in the insulated bag, or wrap it with a heavy towel.

5. Let the yogurt sit undisturbed for eight to 14 hours, or until it is solid. Then, stick it in the refrigerator. It will firm up even more. Homemade yogurt is great plain over fruit, pancakes, or granola cereal (see the recipes in Chapter 3). You can also sweeten it up by stirring in some honey.

FIGURE 5-30: **Pour the milk with the starter yogurt into an insulated mug.**

6. If you prefer thick Greek yogurt, all you need to do is remove some of the liquid— which is called *whey*—from the yogurt you just made. Line a strainer with a coffee filter or paper towel and place it over a bowl. Scoop some yogurt into the strainer and let it sit for about an hour, or until it reaches the desired thickness. Refrigerate the yogurt. Save the whey to use as liquid in baking, or in smoothies for a little extra tang. Since it has the

FIGURE 5-31: **Close the mug and put it in an insulated lunch bag.**

same friendly bacteria as the yogurt itself, you can also use it as starter for fermented ketchup (see the recipe in Chapter 3).

⟫→ WAIT! DON'T EAT ALL OF YOUR YOGURT!

Homemade yogurt will keep for a week or longer. But before it's all gone, be sure to set aside a tablespoon as starter for your next batch!

PROJECT:
PUT TOGETHER A MEAL–SIZED THERMAL COOKER

MATERIALS

Large basket, storage box, cooler, or heavy cloth tote bag (several inches larger all around than your cooking pot)

Thick towels or blankets, padded cloth hot pads, oven mitts, or heavy tote bags

Insulation, such as:

A foil-lined insulated cloth bag just a little bigger than the cooking pot

Aluminum foil or a reflective Mylar "space blanket"

A cloth tote bag big enough to hold the cooking pot

A saucepan, medium casserole dish, or cooking pot with lid—any handle(s) should be short

FIGURE 5-32: **A traditional slow cooker doesn't need electricity—just plenty of padding.**

Cook a dish big enough for your whole family using insulating materials you find around the house. Try different combinations until you find one that works for you. To keep your thermal cooker clean, choose materials that can be sponged off or thrown in the washing machine.

1. Place some padding—like oven mitts or folded towels—in the bottom of the basket. Line the basket with a large towel or blanket, letting the excess hang over the sides. Place another folded towel inside.

2. Then, place the foil-lined bag inside, opened up as wide as possible. If there is any space left in the basket, stuff it with more towels. (You can also place the foil-lined tote bag inside a bigger padded tote bag.)

3. Heat up the food on the stove in a saucepan, following the recipe you are using. Place the cooking pot that you will use for the thermal cooker inside the small tote, and set it on the counter. The bag should be open so you can easily get to the pot.

FIGURE 5-33: **Start with some padding . . .**

FIGURE 5-34: **. . . then add a towel.**

FIGURE 5-35: **Use a foil-lined insulated bag, or line a bag yourself with foil.**

FIGURE 5-36: **Layer one bag inside another for extra padding.**

4. Using oven mitts or hot pads, carefully transfer the hot food from the saucepan to the cooking pot. Replace the lid on the pot, and pull the bag up around it. Use the bag to help you lift the cooking pot up and lower it into the foil-lined bag. Fold the top of both bags over the pot. Cover it with another folded towel. Fold the ends of the big towel over the entire thing and tuck them into the basket.

5. Let your thermal cooker sit for several hours (according to the recipe you are using) until done. Food can stay warm in the cooker for several additional hours without burning.

FIGURE 5-37: **Put the empty dish in the bag before you fill it with hot food. That way, you only need to lift the bag by the handles to transfer the dish to the cooking basket.**

FIGURE 5-38: **Keep everything covered while it's cooking.**

RECIPE: THERMAL COOKER LENTILS AND RICE

🥄 MATERIALS

1 cup lentils (any kind will work, but red and yellow cook faster than green)

1 cup white rice

4 cups water

Optional: **Sliced carrots, chopped spinach, onions, or other vegetables**

Salt, pepper and/or other spices to taste (try ½ teaspoon cinnamon or curry powder)

FIGURE 5-39: **A hearty dish that cooks itself**

Lentils are sold dried and resemble split peas. Cook them together with rice for a hearty vegetarian main dish.

1. *Optional:* Soak the lentils in cold water for eight hours to shorten the cooking time. It will also make them easier to digest.

2. Prepare the thermal cooker, as described in the previous project.

3. Bring water to boil in a covered saucepan on the stove.

4. Add the rice and lentils. If you're adding vegetables, throw them in, too. Bring the pot back to a boil; then lower the heat and let simmer for 5 or 10 minutes.

FIGURE 5-40: **Sliced carrots and cinnamon add flavor to lentils and rice.**

5. Carefully transfer the rice, lentils, and water to the cooking pot. Lower the pot into the thermal cooker using the cloth tote bag. Let it finish cooking for two or three hours, or until the lentils are soft, but not mushy.

6. Before serving, add salt and any other seasonings you choose.

DiY OUTDOOR STOVE

FIGURE 5-41: **A DIY camping stove lets you cook outdoors without a barbeque or campfire.**

Campfires are fun. They're also easy to cook with. Just stick a hot dog on a stick and hold it over the fire, or wrap your fish up in some aluminum foil and throw it on the coals. But even in areas where finding firewood isn't a problem, sometimes it's best to bring your own stove and fuel. Popular camping spots are often picked clean of dead branches and logs, and in dry weather there's the danger that a campfire will spread to the surrounding vegetation. A camping stove that's light enough to carry around is handy for these situations—or if you just want to cook a quick meal in your backyard.

PROJECT: TiN CAN COOKER

MATERIALS

Several small tin cans (tuna cans, or candy tins with a lid)

Rotary can opener (the kind you use to open a can of cat food)

Strips of corrugated cardboard, slightly wider than the cans are tall

Unscented white candles or paraffin wax blocks

Double boiler to melt wax (or a medium saucepan and a slightly larger saucepan)

Stove (preferably electric) or hot plate

Large tin can (#10 institutional size is best)

Church key–type can opener (the kind that makes a triangle-shaped hole in the can)

Pliers

Optional: Tin snips

Heavy work gloves

Metal tongs

Hot pads or oven mitts

Brick or metal pot lid

Matches or a lighter

The Tin Can Cooker is a home-made outdoor stove that really works. Generations of Scouts have learned to make and use "Buddy Burners" for camping and emergencies. Instead of wood or liquid fuel, the Tin Can Cooker burns a small can of wax and cardboard. A larger can fits over it and holds your cooking pot. Each can of wax should burn an hour or more. Make several so you always have one on hand.

FIGURE 5-42: **A portable stove made from two recycled cans and some old candles**

SAFETY TiPS FOR TiN CAN COOKERY

Tin Can Cookers use an open flame. They can be dangerous, so be sure to follow these safety rules!

- Kids need adult supervision to make and use a Tin Can Cooker.

- Wear heavy gloves when handling cut metal edges.

- Always keep an eye on melting wax and on the lit cooker. Do not leave them unattended.

- Do not melt wax in the microwave or directly on the stove. Always use a double boiler (or make one from two saucepans).

- Always use hot pads or oven mitts to handle hot pots.

- The entire Tin Can Cooker gets hot, so never touch it with bare hands while it is lit! Use nonflammable oven mitts or metal tongs to adjust the cooking can or the wax burner when they are in use.

- Make sure to use your Tin Can Cooker on a level, fireproof surface, such as a flat paving stone. Watch out for hot liquid wax in the burner—avoid spilling it.

- Do not pour water on burning wax. Use a brick or a metal lid to cover the wax can and smother the flames.

- Tin Can Cookers can blacken cooking utensils with soot from the burning wax, so don't use them with your best pots and pans. Pick up a camping skillet and spatula or small, sturdy garage sale pots and pans and save them just for outdoor cooking.

Tip Be sure to clean the cans you're using to make your cooker and wax burner carefully before using. Remove the top completely and throw it away. Also, remove the label and any glue. Avoid using cans with plastic coatings.

1. Take each tuna can and curl up strips of cardboard to fit inside loosely, and make sure you leave enough space for wax. Or cut short strips and arrange them to form a star that divides the can into pie-shaped slices.

Set the prepared cans on a protected surface, such as a cookie sheet covered in newspaper or aluminum foil, and place it close to where you'll be melting the wax.

FIGURE 5-43: **Cardboard acts as a wick to get the wax burning across the top of the can.**

2. Break up the wax or old candles, if needed, to fit into the top pan of the double boiler. Fill the bottom pan about halfway with water. Fit the top pan into it so it sits above the water. Turn the stove on medium to heat the water to a steady boil. Let the wax melt until it is completely liquid.

3. Carefully pour the melted wax into the cans. Leave the top edge of the cardboard sticking up above the level of the wax. If your wax starts to harden as you're pouring it, just reheat it to melt it again. Let cool until hardened.

4. Hold the large can with the opening pointing away from you. Take the church key can opener (the kind you use to punch holes in a juice can) and make five or six openings around the side, about an inch apart.

FIGURE 5-44: **If you don't have a double boiler, sit a medium-sized saucepan over a larger saucepan of boiling water.**

FIGURE 5-45: **If the wax starts to harden as you're pouring it, just reheat it to melt it again.**

5. Turn the large can around so the opening is toward you. You have two options for letting air in at the bottom to feed the flames. If you don't have tin snips, use the can opener to punch holes all around the bottom of the can, about 3 inches apart. Then take a pair of pliers and fold over the sharp points of metal created by each hole and squeeze them flat. If you do have tin snips, create a flap that acts like a fireplace damper to control the airflow. Put on heavy work gloves to protect your hands from sharp metal edges. Use the tin snips to cut two slits into the side, about 3 inches long and 3 inches apart. Then, grab the edge of the can between the slits and bend it outward to make a flap that can open and close. The more oxygen you let in, the higher the flames will be.

6. To use your stove, place the wax can on a flat nonflammable surface. Light the cardboard strips, like lighting the wick of a candle. You want the flame to cover the entire top of the can. When the wax can is lit, adjust the flap on the large can (if there is one) and place

FIGURE 5-46: **Use a can opener to punch air holes near the top of the "stove" can.**

FIGURE 5-47: **Use pliers to fold over the sharp points to avoid cuts.**

FIGURE 5-48: **Wear heavy gloves when cutting the metal can to protect against sharp edges.**

it over the smaller can so the holes on the top are facing away from you. (This keeps smoke from blowing in your face.) Safety note: Remember, the entire tin can stove gets hot, so never touch it with bare hands while it is lit! Use nonflammable oven mitts and a pair of pliers if you need to adjust the flap while the stove is in use.

RECIPE: GRILLED CHEESE

1. Butter one slice of bread and put it buttered side down on a plate. Lay on slices of cheese. Place another slice of bread on top. Butter the top.

2. Place the grilled cheese sandwich in a frying pan. Next, carefully place the skillet on the tin can cooker. After a few minutes, use a spatula to flip the sandwich over—it should be toasty on both sides, with the cheese melted inside.

3. *Optional:* Shake on some pizza spices and add thin slices of tomato to your sandwich. Make sure to put cheese above and below the tomato to hold the sandwich together.

MATERIALS

2 slices bread

Butter

Sliced American or other cheese

FIGURE 5-49: **Grilled cheese—simple and perfect.**

RECIPE:
EGG-IN-A-HOLE

MATERIALS

Butter

1 slice bread

1 egg

FIGURE 5-50: **Make your egg and your toast at the same time with this favorite campfire recipe.**

1. Use a small juice glass or round cookie cutter to cut a hole out of the center of the bread. Save the round piece of bread to toast.

FIGURE 5-51: **Use a juice glass or cookie cutter to make a hole in the bread.**

2. Melt some butter in a frying pan and spread it around by tilting the pan. Lay the bread in the pan for a minute to soak up some of the butter. Then, flip the bread over to let melted butter soak in on the other side. (The circle of bread you cut out to make the hole can be toasted the same way.)

3. Crack the egg with a fork or on the edge of a bowl—not on the edge of the hot pan—and carefully drop it into the hole in the bread. Try not to break the yolk.

FIGURE 5-52: Crack the egg away from the flame, and watch out for the hot can as you pour it into the pan.

4. Let the egg cook until all of the clear part has turned white except the very top. Use a spatula to flip over the bread and egg. Also flip over the circle of bread if you are cooking them together.

FIGURE 5-53: **This egg is almost ready to flip.**

5. Let the egg cook another minute or two until the yolk is firm. Transfer it to a plate and season with salt and pepper to taste.

MORE ABOUT ALTERNATIVE COOKING METHODS

Cooking for Geeks: Real Science, Great Hacks, and Good Food by Jeff Potter (O'Reilly, 2010)

Solar Cooker at CantinaWest (http://www.solarcooker-at-cantinawest.com)

How Solar Cooking Works by Julia Layton (http://science.howstuffworks.com/environmental/green-science/solar-cooking.htm)

Solar Cookers International Network Wiki (http://solarcooking.org)

Wonderbag (http://www.wonderbagworld.com)

Index

NUMBERS

2D axes, 18

3 Digital Cooks, 36

3D food printing, 1, 14–15. *See also* hydraulic LEGO 3D food printer

3D purple house, 19

A

acids and bases, 44

agar, 39

agar noodles, 45–47. *See also* noodles

Agar-Agar Raindrop Cake, 50–51

alfalfa seeds, sprouting, 109–111

alternative cookers, 135, 169

Amazing Food Made Easy, 77

Appert, Nicholas, 80

applesauce, making, 103–104

Applesauce Cake in a Mug, 101–102

aquafaba, 56, 98

aquaponic jar, making, 121–127

The Aquaponic Source, 132

aquarium, growing plants in, 122

ArtBot, ButterBot variation of, 9

avocado tree, planting, 118–119

B

bacteria

killing off, 135

preventing growth of, 90

Baked Foam Meringue Cookies, 53–56. *See also* foams

baked goods, cooking temperature for, 135

baking powder

invention of, 85

using in pancake mix, 86

baking soda

in Fizzy Watermelon Lemonade, 63, 65

pH of, 44

using in pancake mix, 86

bananas, using in Dry Ice Sorbet, 73

bases and acids, 44

basil

growing in water, 125

growing indoors, 115

battery, using with Motorized ButterBot, 9

Battle Creek cereals, 82

beets, regrowing leaves on, 116

Birdseye, Clarence, 80

bog butter, 3

boiling water, 135

Bonaparte, Napoleon, 80

Bordessa, Kris, 105

Born, Lily, xiv–xv, xvi, 36

bouncy spheres, 39

brainstorming, xv

breakfast. *See* cold breakfast

brown sugar, using in Applesauce Cake in a Mug, 101–102. *See also* sugar

bubbles. *See* foams

Build an Indoor Worm Bin, 129–131

building materials, xx

butter

    in bottle, 7

    flavoring, 6

    making in jar, 4–6

    storing, 6

butter maker, 3

ButterBot, making, 8–13

buttermilk, 3–6

C

Caleb, James, 82

camping stove, 160

candy thermometer, using with yogurt maker, 153

canning, development of, 80

carbon dioxide gas, using to create foam, 52

cardboard solar oven, building, 138–145

carrots, regrowing leaves on, 116

Cartesian plane, 18

celery, growing indoors, 114

Celsius (°C), xxii

chemical cuisine resources, 77

chemicals, adding to dishes, xiii

chemistry, of crystals, 66

chickpeas

    in hummus, 97–99

    protein content of, 56

chlorophyll, 112

chocolate bar, using with Marshmallow Sunpies, 61

chocolate cake, solar, 149–150

chocolate hazelnut spread, 33

chork (chopsticks and fork), 2

cider vinegar, using in ketchup, 91. *See also* vinegar; white vinegar

citric acid, pH of, 44

cocoa powder, using in Solar Chocolate Cake, 149

cold breakfast, history of, 82

Collins, Sarah, 151

compost jar, using, 128

compound butter, making, 6

condiments and spreads, 89–90

Consider the Fork, 36

cookies, baking in cardboard oven, 133. *See also* meringue cookies; oatmeal cookies

cooking

    equipment, xix

    "off the grid," 134

    from scratch, 81, 105

    with sun, 136

    tips, xxii

    tools, 36

Cooking for Geeks: Real Science, Great Hacks, and Good Food, 169

cooking pans, solar, 139

coordinates, specifying on graphs, 17

crafts, xx

craft-stick weight, using with ButterBot, 12

cream
  separating in ButterBot, 13
  using to make butter, 4, 8
cream of tartar
  in Baked Foam Meringue
    Cookies, 54
  pH of, 44
Creative Commons license, xii
Crunchy Granola, 83–84
crystals, chemistry of, 66
cucumbers, making pickles
  with, 95–96
Culinary Reactions, 77
cup. *See* Kangaroo Cups

D

dasher, using to make butter, 3
Dazey, Nathan, 3
DC motor, using with Motorized
  ButterBot, 9, 11
DiJusto, Patrick, 105
dill, using for pickles, 95–96
*Do the Rot Thing: A Teacher's Guide
  to Compost Activities*, 132
*Don't Throw It, Grow It! 68
  Windowsill Plants from Kitchen
  Scraps*, 132
drawings, plotting on graphs, 17–19
Drinking-Straw Noodles, 48
dry ice
  making water "boil" with, 70
  safety rules, 71
  sorbet, 72–75
  storing, 71
  temperature of, 70
  using, 71

drying technique, using, 66
Durand, Peter, 80

E

edible plants, growing, 132. *See also*
  plants
effervescence, explained, 52
egg substitute, aquafaba as, 98
Egg-in-a-Hole, 167–168
eggless meringue, 56
eggs
  cooking temperature, 135
  in Pancake Mix Pancakes, 87
egg-separating device, using, 57
equipment, xix, xxi–xxii
experimenting, xvi

F

Fahrenheit (°F), xxii
fan, using as motor, 11
fan base, building, xii
fats, melting temperature, 135
fermentation, 90, 92
fish
  choosing for aquaponics, 122
  feeding, 126
  moving, 125
fish tanks, home ecology
  aquaponics kit for, 132
Fizzy Watermelon
  Lemonade, 63–65
flavorings, adding to
  marshmallows, 60
flour
  in Applesauce Cake in a
    Mug, 101–102

in Solar Chocolate Cake, 149

using in pancake mix, 86

foams. *See also* Baked Foam

Meringue Cookies

creating, 52

textures, 52

Food and Drug Administration, 89

food coloring, adding to Rock

Candy Sticks, 68

food printing. *See* 3D food printing

Food Rules: An Eater's Manual, 105

foods. *See also* Pure Food and Drug

Act

pre-prepared, 80–81

refrigerating, 90

fork

and chopsticks (chork), 2

using with Motorized

ButterBot, 12

French fries, oven-baking, 93–94

fresh vegetables, benefits, 108. *See*

also vegetables

frostbite injuries from dry ice,

avoiding, 71, 74

frozen desserts, making, 70–74

frozen foods, popularity of, 80

frozen fruit puree, making, 72–75

fruit, putting in raindrop cake, 51

fruit puree, making, 72–75

G

garlic, growing indoors, 113–117

gel noodles, 45

gelatin, 39

gelatin dots, 41–44

gelatin marshmallows, 58–60

gels

causing to glow, 49

using, 39

glowing gels, making, 49

graham cracker, using in

Marshmallow Sunpies, 61

granola

invention of, 82

recipe, 83–84

grape juice dots, changing color

of, 44

graphs, plotting drawings on, 17–19

gravel, using in aquarium, 124

Grilled Cheese, 166

Grow Instant Indoor Veggies from

Kitchen Cuttings, 113–117

grow lights, using, 114

Grow Sprouts in a Jar, 109–112

Guacamole Dip with Tomato, 120.

See also tomato-flavored gel

noodles

H

Halloran, Amy, 86, 105

Harvard University Science and

Cooking videos, 77

heavy cream, using to make

butter, 4, 8

Heinz, Henry John, 89–90

help, asking for, xvi

herbs, growing indoors, 113–117

Homemade Hummus, 97–99

Homemade Whipped Gelatin

Marshmallows, 58–60

honey

in granola, 83–84

in Tangy Fermented Ketchup
   from Scratch, 91
Horsford, Eben Norton, 85
"hot box," 136
hot food, handling, 156
hot mixtures, safety tips for
   working with, 38
household materials, xx
How Solar Cooking Works, 169
hummus, making, 97–99
hydraulic LEGO 3D food printer.
   See also 3D food printing; LEGOs
   arch and bow pieces, 25, 27
   brick with knobs, 24
   chocolate hazelnut spread, 33
   crossbeam, 27–28
   food tube arm, 29–30
   frame, 25–28
   frosting and food tube, 21
   guide for food tube, 28
   holder for syringe, 25
   hydraulic systems, 31–32
   inverted roof tiles, 25
   ledge, 26–27
   materials, 20
   moving parts, 21–22
   print bed and rails, 23–25
   printing, 33–35
   squeezing frosting, 22
   straw, 34
   syringe behind arch, 32
   syringe holder, 26
   syringe on ledge, 32
   syringe plunger, 23
   tiles, 25
   towers, 26

windows, 28
hydrocolloids, 39
hydrogen ion (H+), formation of, 44
hydroponics, 121

I
ice cream, making, 76
ice sorbet, making, 75–76
ideas
   brainstorming, xv
   protecting, xi
indoor veggies, growing, 113–117
indoor worm bin, building, 129–131
Industrial Revolution, 2
ingredients, xviii
inventions
   process, xiv–xvi
   taking to market, xi
iterating, xvi

J
Jefferson, Thomas, 93
juicy gelatin dots, 41–44

K
Kangaroo Cups, xiv–xv , 36
Kellogg, John Harvey, 82
ketchup
   history of, 89
   making from scratch, 91–92
KidsGardening.org, 132
kitchen
   safety tips, 85
   staples, xviii–xx

L
launching, xvi

Layton, Julia, 169
leafy greens, growing
    indoors, 113–117
leeks, growing indoors, 114
LEGO PancakeBot, xi–xii
LEGOs. *See also* hydraulic LEGO
    3D food printer
    bricks, 22
    building slots, 23
    building with, 22–23
    Digital Designer, 23
    extra pieces, 23
    graphing, 19
    knobs, 22
    plates, 22
    set and baseplate, 23
    set and element ID numbers, 22
    shapes, 22
    sizes of pieces, 22
    tiles, 22
lemon juice
    in applesauce, 103–104
    in hummus, 97–99
    pH of, 44
lemonade, making with
    watermelon, 63–65
lentils and rice, 158–159
Lersch, Martin, 77
licensing, considering, xi–xii
Löf, George, 136–137

M

Maillard reaction, 135
Make an Aquaponic Jar, 121–127
mangos, using in Dry Ice Sorbet, 73
marketing, xvi

Marshmallow Sunpies, 61–62
marshmallows, 58–60
materials, xxi–xxii
    agar noodles, 45
    Agar-Agar Raindrop Cake, 50
    Applesauce Cake in a
        Mug, 101–102
    Baked Foam Meringue
        Cookies, 53
    Build an Indoor Worm
        Bin, 129–131
    butter in jar, 4
    ButterBot, 8
    cardboard solar oven, 138
    Dry Ice Sorbet, 72
    Egg-in-a-Hole, 167–168
    Fizzy Watermelon
        Lemonade, 63
    gelatin dots, 41–44
    Grilled Cheese, 166
    Grow Instant Indoor
        Veggies from Kitchen
        Cuttings, 113–117
    Grow Sprouts in a Jar, 109–112
    Guacamole Dip with
        Tomato, 120
    Homemade
        Applesauce, 103–104
    Homemade Hummus, 97–99
    Homemade Whipped Gelatin
        Marshmallows, 58
    ice sorbet, 75
    Make an Aquaponic Jar, 121–127
    Marshmallow Sunpies, 61
    Meal-Sized Thermal
        Cooker, 155–157

Oven-Baked French Fries, 93

Pancake Mix and
    Pancakes, 86–87

Plant an (Almost) Instant
    Avocado Tree, 118–119

plotting drawings on graphs, 17

Rock Candy Sticks, 67–69

Self-Contained Countertop
    Yogurt-Maker, 152–154

Solar Chocolate Cake, 149–150

Solar Nachos, 147

Solar Oatmeal Cookies, 148

Sweet Refrigerator
    Pickles, 95–96

Tangy Fermented Ketchup
    from Scratch, 91

Thermal Cooker Lentils and
    Rice, 158–159

Tin Can Cooker, 161–165

Use a Compost Jar to Turn
    Veggie Scraps into Soil, 128

Mathe, Moshy, 151

Meal-Sized Thermal
    Cooker, 155–157

measurements, xxii

meat, browning, 135

meringue cookies, 53–56. *See also*
    cookies; foams; oatmeal cookies

metric units, xxii

microwave revolution, 100

mint tea, adding to gel noodles, 39

mizu shingen mochi, 50

model, creating, xv

modernist cuisine
    bouncy spheres, 39
    chemical enemies, 40
    gels, 39
    goal of, 38
    rubbery noodles, 39

molecular gastronomy. *See*
    modernist cuisine

Motorized ButterBot, making, 8–13

N

Nabisco Shredded Wheat, 82

nachos, solar, 147

The New Bread Basket, 105

nitrogen gas, transforming into
    liquid, 70

noodles, 39, 48. *See also* agar
    noodles

notes, keeping, xvi

nuts and seeds, using in
    granola, 83–84

O

oatmeal cookies, solar, 148. *See also*
    cookies; meringue cookies

oats, using in granola, 83–84

observing, xv

*Off the Shelf: Homemade Alternatives
    to the Condiments, Toppings, and
    Snacks You Love*, 105

oil splatters, avoiding burns from, 93

*The Omnivore's Dilemma: The Secrets
    behind What You Eat*, 105

onions, dicing for pickles, 95–96

onion-y plants, growing
    indoors, 113–117

"open source," xii

outdoor stove, 160

Oven-Baked French Fries, 93–94

P

pancake art, making, 16

Pancake Mix and Pancakes, 86–88

Pancake Painter, 16–19

PancakeBot, xi–xii, 14–15, 22, 36

Pasteur, Louis, 80

patent application, filing, xi

pea seeds, sprouting, 109–111

Perky, Henry, 82

Peterson, Deborah, 132

pH, explained, 44

photos, taking, xvi

photosynthesis, 112

pickling, 90, 95–96

pineapple and gelatin, avoiding, 40

Plant an (Almost) Instant Avocado
    Tree, 118–119

plants. *See also* edible plants
        growing in water, 121–122
        turning green in sunlight, 112

plotting drawings on graphs, 17–19

Pollan, Michael, 105

Post, C. W., 82

potassium bitartrate, pH of, 44

potatoes, using for French
    fries, 93–94

Potter, Jeff, 169

powdered sugar, 58. *See also* sugar

pre-prepared foods, 80–81

printing materials,
    preparing, 33–35

processed food, 80

projects
        agar noodles, 45–47
        Build a Cardboard Solar
            Oven, 138
        Build an Indoor Worm
            Bin, 129–131
        ButterBot, 8–13
        Grow Instant Indoor
            Veggies from Kitchen
            Cuttings, 113–117
        Grow Sprouts in a Jar, 109–112
        handmade butter in jar, 4–6
        ice sorbet, 75–76
        juicy gelatin dots, 41–44
        Make an Aquaponic Jar, 121–127
        Meal-Sized Thermal
            Cooker, 155–157
        Plant an (Almost) Instant
            Avocado Tree, 118–119
        plotting drawings on
            graphs, 17–19
        Rock Candy Sticks, 67–69
        Self-Contained Countertop
            Yogurt-Maker, 152–154
        Tin Can Cooker, 161–165
        Use a Compost Jar to Turn
            Veggie Scraps into Soil, 128

prototyping, xv

protractor, using with cardboard
    solar oven, 143

Pure Food and Drug Act, 89. *See
    also* foods

Q

Quellen Field, Simon, 77

Quick Freeze Machine, 80

quinine, 49

R

radish seeds, sprouting, 109–111

Raindrop Cake, 50
raisins, using in granola, 83–84
recipes. *See also* solar recipes
    Agar-Agar Raindrop
      Cake, 50–51
    Applesauce Cake in a
      Mug, 101–102
    Baked Foam Meringue
      Cookies, 53–56
    Crunchy Granola, 83–84
    Dry Ice Sorbet, 72–75
    Egg-in-a-Hole, 167–168
    Fizzy Watermelon
      Lemonade, 63–65
    Grilled Cheese, 166
    Guacamole Dip with
      Tomato, 120
    Homemade
      Applesauce, 103–104
    Homemade Hummus, 97–99
    Homemade Whipped Gelatin
      Marshmallows, 58–60
    Marshmallow Sunpies, 61–62
    Oven-Baked French
      Fries, 93–94
    Pancake Mix and
      Pancakes, 86–88
    Solar Chocolate Cake, 149–150
    Solar Nachos, 147
    Solar Oatmeal Cookies, 148
    Sweet Refrigerator
      Pickles, 95–96
    Tangy Fermented Ketchup
      from Scratch, 91–92
    Thermal Cooker Lentils and
      Rice, 158–159

red wiggler worms, using in worm
    bin, 129
refrigerating homemade food, 90
rice and lentils, 158–159
rock candy crystals, growing, 66
Rock Candy Sticks, 67–69
rolled oats, using in granola, 83–84
romaine lettuce, growing
    indoors, 114–115
royalty payments, receiving, xi–xii
rubbery noodles, 39
Rumford Baking Powder, 86
Rumford Chemical Works, 85

S

safety notes, xxiii–xxiv
    aquaponics, 123
    cooking and cutting, 85
    dry ice, 71
    handling hot food, 156
    hot mixtures, 38
    sticky mixtures, 38
    temperatures, 135
de Saussure, Horace-Bénédict, 136
scallions, growing indoors, 114
seeds and nuts, using in
    granola, 83–84
self-contained cooking, 151–154
Self-Contained Countertop
    Yogurt-Maker, 152–154
Selsam, Millicent, 132
shredded wheat, 82
silicone-rubber tubing, 39, 48
slow cookers, 151
S'mores, making, 60
snails, keeping in aquaponic jar, 124

sodium bicarbonate, pH of, 44

soft matter, 39

Solar Chocolate Cake, 149–150

solar cookers, 169

solar cooking pans, 139

Solar Nachos, 147

Solar Oatmeal Cookies, 148

solar ovens

    building from

      cardboard, 138–145

    designs, 137

    invention of, 136

solar recipes, 146. *See also* recipes

sorbet, making, 72–76

Spencer, Percy, 100

spreads and condiments, 89–90

Sprout People, 132

sprouts, growing in jar, 109–111

sticky mixtures, safety tips for

  working with, 38

story, telling, xii

sugar. *See also* brown sugar;

  powdered sugar

  in Agar-Agar Raindrop

    Cake, 50

  in Baked Foam Meringue

    Cookies, 53

  caramelizing, 135

  in Dry Ice Sorbet, 72

  in Homemade Whipped

    Gelatin Marshmallows, 58

  in Rock Candy Sticks, 67

  in Sweet Refrigerator

    Pickles, 95

  using with Dry Ice Sorbet, 73

sun, cooking with, 136

supersaturated solution, 66

supplies, xix, xxi–xxii

Sweet Refrigerator Pickles, 95–96

T

tahini, using in hummus, 97–99

Tangy Fermented Ketchup from

  Scratch, 91–92

Telkes, Mária, 136–137

temperatures

  and food safety, 135

  for ingredients, 135

  for killing off bacteria, 135

  measurement in degrees, xxii

  and times, 134

testing prototypes, xvi

Texture: A Hydrocolloid Recipe

  Collection, 77

Thermal Cooker Lentils and

  Rice, 158–159

thermal cookers, 151, 155–157

This Is What You Just Put in Your

  Mouth? 105

Thomas, Gerry, 80

Tin Can Cooker, 161–165

tin can, development of, 80

tips, xxii

tomato paste, getting out of cans, 92

tomato-flavored gel noodles, 45. *See*

  also Guacamole Dip with Tomato

tonic water, using with gel, 49

tortilla chips, using for Solar

  Nachos, 147

TV dinner, invention of, 80

U

Umbroiler, 136–137
Use a Compost Jar to Turn Veggie
Scraps into Soil, 128
U.S. units, xxii

V

Valenzuela, Miguel, 14
vanilla extract
in Baked Foam Meringue
Cookies, 53
in Crunchy Granola, 83
in Homemade Whipped
Gelatin Marshmallows, 58
in Solar Chocolate Cake, 149
in Solar Oatmeal Cookies, 148
vegetables. *See also* fresh
vegetables
cooking temperature, 135
growing in aquaponic jar, 122
growing indoors, 113–117
packaging, 80, 112
veggie scraps, turning into soil, 128
Vibrating ArtBot, 9
videos, taking, xvi
vinegar. *See also* cider vinegar;
white vinegar
pH of, 44
using to kill bacteria, 90
volcano. *See* foams

W

water
boiling, 135
growing plants in, 121

water bottle, using to separate
eggs, 57
water cake, 50
watermelon lemonade, 63–65
wax, melting, 163
websites
alternative cookers, 169
Born, Lily, xv
chemical cuisine resources, 77
cooking from scratch, 105
cooking tools, 36
Creative Commons license, xii
growing edible plants, 132
Pancake Painter, 19
United States Patent Office, xi
weight, using with ButterBot, 11–12
white vinegar, using for
pickles, 95–96. *See also* cider
vinegar; vinegar
Wilson, Bee, 36
Wohlt, Goose, 56
Wonderbag, 151, 169
worm bin, building, 129–131

X

X-Y plane, 18

Y

Y axis, 19
yogurt
in pancake mix pancakes, 87
whey from, 92
yogurt-maker, 152–154

Z

Z axis, 18

Also available from Make:

ReMaking History, Volume 1: Early Makers

By William Gurstelle

Learn about inventors and inventions from the distant past that have shaped the world we live in today. Then, make those inventions yourself with step-by-step instructions that show you how. Bring history to life!

ISBN: 978-1680450606 | $19.99